SUPER
CRUNCHERS

Also by Ian Ayres

---

*Responsive Regulation: Transcending the Deregulation
Debate* (1992)

Studies in Contract Law (1997)

*Voting with Dollars: A New Paradigm for Campaign
Finance* (2002)

*Pervasive Prejudice?: Unconventional Evidence of Race
and Gender Discrimination* (2002)

*Why Not?: How to Use Everyday Ingenuity to Solve
Problems Big and Small* (2003)

*Straightforward: How to Mobilize Heterosexual Support
for Gay Rights* (2005)

Insincere Promises: The Law of Misrepresented Intent
(2005)

Optional Law: The Structure of Legal Entitlements
(2005)

IAN AYRES

SUPER
CRUNCHERS

How Anything Can Be Predicted

John Murray

First published in Great Britain in 2007 by John Murray (Publishers)
An Hachette Livre UK company

1

© Ian Ayres, 2007

A CIP catalogue record for this title is available from the British Library

Hardback ISBN 978-0-7195-6463-5
Trade Paperback ISBN 978-0-7195-6464-2

Printed and bound by Clays Ltd, St Ives plc

John Murray policy is to use papers that are natural, renewable and recyclable
products and made from wood grown in sustainable forests. The logging and
manufacturing processes are expected to conform to the environmental
regulations of the country of origin.

John Murray (Publishers)
338 Euston Road
London NW1 3BH

www.johnmurray.co.uk

For John Donohue and Peter Siegelman,
who were there when I needed them most

CONTENTS

SUPER
CRUNCHERS

INTRODUCTION

The Rise of the
Super Crunchers

Orley Ashenfelter really loves wine: "When a good red wine ages," he says, "something quite magical happens." Yet Orley isn't just obsessed with how wine tastes. He wants to know about the forces behind great and not-so-great wines.

"When you buy a good red wine," he says, "you're always making an investment, in the sense that it's probably going to be better later. And what you'd like to know is not what it's worth now, but what it's going to be worth in the future. Even if you're not going to sell it—even if you're going to drink it. If you want to know, 'How much pleasure am I going to get by delaying my gratification?' That's an endlessly fascinating topic." It's a topic that has consumed a fair amount of his life for the last twenty-five years.

In his day job, Orley crunches numbers. He uses statistics to extract hidden information from large datasets. As an economist at Princeton, he's looked at the wages of identical twins to estimate the impact of an extra year of school. He's estimated how much states value a statistical life by looking at differences in speed limits. For years he edited the leading economics journal in the United States, the *American Economic Review*.

Ashenfelter is a tall man with a bushy mane of white hair and a booming, friendly voice that tends to dominate a room. No milquetoast he. He's the kind of guy who would quickly disabuse you of any stereotype you might have that number crunchers are meek, retiring souls. I've seen Orley stride around a classroom, gutting the reasoning behind a seminar paper with affable exuberance. When he starts out his remarks with over-the-top praise, watch out.

What's really gotten Orley in trouble is crunching numbers to assess the quality of Bordeaux wines. Instead of using the "swishing and spitting" approach of wine gurus like Robert Parker, Orley has used statistics to find out what characteristics of vintage are associated with higher or lower auction prices.

"It's really a no-brainer," he said. "Wine is an agricultural product dramatically affected by the weather from year to year." Using decades of weather data from France's Bordeaux region, Orley found that low levels of harvest rain and high average summer temperatures produce the greatest wines. The statistical fit on data from 1952 through 1980 was remarkably tight for the red wines of Burgundy as well as Bordeaux.

Bordeaux are best when the grapes are ripe and their juice is concentrated. In years when the summer is particularly hot, grapes get ripe, which lowers their acidity. And, in years when there is below-average rainfall, the fruit gets concentrated. So it's in the hot and dry years that you tend to get the legendary vintages. Ripe grapes make supple (low-acid) wines. Concentrated grapes make full-bodied wines.

He's had the temerity to reduce his theory to a formula:

Wine quality = 12.145 + 0.00117 winter rainfall + 0.0614 average growing season temperature − 0.00386 harvest rainfall

That's right. By plugging the weather statistics for any year into this equation, Ashenfelter can predict the general quality of any vintage. With a slightly fancier equation, he can make more precise predictions for the vintage quality at more than 100 Châteaus. "It may seem a bit mathematical," he acknowledges, "but this is exactly the way the French ranked their vineyards back in the famous 1855 classifications."

Traditional wine critics have not embraced Ashenfelter's data-driven predictions. Britain's *Wine* magazine said "the formula's self-evident silliness invite[s] disrespect." William Sokolin, a New York wine merchant, said the Bordeaux wine industry's view of Ashenfelter's work ranges "somewhere between violent and hysterical." At times, he's been scorned by trade members. When Ashenfelter gave a wine presentation at Christie's Wine Department, dealers in the back openly hissed at his presentation.

Maybe the world's most influential wine writer (and publisher of *The Wine Advocate*), Robert Parker, colorfully called Ashenfelter "an absolute total sham." Even though Ashenfelter is one of the most respected quantitative economists in the world, to Parker his approach "is really a Neanderthal way of looking at wine. It's so absurd as to be laughable." Parker dismisses the possibility that a mathematical equation could help identify wines that actually taste good: "I'd hate to be invited to his house to drink wine."

Parker says Ashenfelter "is rather like a movie critic who never goes to see the movie but tells you how good it is based on the actors and the director."

Parker has a point. Just as it's more accurate to *see* the movie, shouldn't it be more accurate to actually *taste* the wine? There's just one catch: for months and months there is no wine to taste. Bordeaux and Burgundies spend eighteen to twenty-four months in oak casks before they are set aside for aging in bottles. Experts, like Parker, have to wait four months just to have a first taste after the wine is placed in barrels. And even then it's a rather foul, fermenting mixture. It's not clear that tasting this undrinkable early wine gives tasters very accurate information about the wine's future quality. For example, Bruce Kaiser, former director of the wine department at auctioneer Butterfield & Butterfield, has claimed, "Young wines are changing so quickly that no one, and I mean no one, can really evaluate a vintage at all accurately by taste until it is at least ten years old, perhaps older."

In sharp contrast, Orley crunches numbers to find the historical relationship between weather and price. That's how he found out that each centimeter of winter rain tends to raise the expected price 0.00117 dollars. Of course, it's only a tendency. But number crunching lets Orley predict the future quality of a vintage as soon as the grapes are harvested—months before even the first barrel taste and years before the wine is sold. In a world where wine futures are actively traded, Ashenfelter's predictions give wine collectors a huge leg up on the competition.

Ashenfelter started publishing his predictions in the late eighties in a semiannual newsletter called *Liquid Assets*. He advertised the newsletter at first with small ads in *Wine Spectator,* and slowly built a base of about 600 subscribers. The subscribers were a geographically diverse lot of millionaires and wine buffs—mostly confined to the small group of wine collectors who were comfortable with econometric techniques. The newsletter's subscription base was just a fraction of the 30,000 people who plunked down $30 a year for Robert Parker's newsletter *The Wine Advocate.*

Ashenfelter's ideas reached a much larger audience in early 1990 when the *New York Times* published a front-page article about his new prediction machine. Orley openly criticized Parker's assessment of the 1986 Bordeaux. Parker had rated the '86s as "very good and sometimes exceptional." Ashenfelter disagreed. He felt that the below-average growing season temperature and the above-average harvest rainfall doomed this vintage to mediocrity.

Yet the real bombshell in the article concerned Orley's prediction about the 1989 Bordeaux. While these wines were barely three months in the cask and had yet to even be tasted by critics, Orley predicted that they would be "the wine of the century." He guaranteed that they would be "stunningly good." On his scale, if the great 1961 Bordeaux were 100, then the 1989 Bordeaux were a whopping 149. Orley brazenly predicted that they would "sell for as high a price as any wine made in the last thirty-five years."

The wine critics were incensed. Parker now characterized Ashenfelter's quantitative estimates as "ludicrous and absurd." Sokolin said the reaction was a mixture of "fury and fear. He's really upset a lot of people." Within a few years *Wine Spectator* refused to publish any more ads for his (or any other) newsletter.

The traditional experts circled the wagons and tried to discredit both Orley and his methodology. His methodology was flawed, they said, because it couldn't predict the exact future price. For example, Thomas Matthews, the tasting director of *Wine Spectator,* complained that Ashenfelter's price predictions "were exactly true only three times in the twenty-seven vintages." Even though Orley's "formula was specifically designed to fit price data" his "predicted prices are both under and over the actual prices." However, to statisticians (and to anyone else who thinks about it for a moment), having predictions that are sometimes high and sometimes low is a good thing; it's the sign of an unbiased estimate. In fact, Orley showed that Parker's initial ratings of vintages had been systematically biased upward. More often than not Parker had to downgrade his initial rankings.

In 1990 Orley went even further out on a limb. After declaring 1989 the vintage of the century, the data told him that 1990 was going to be even better. And he said so. In retrospect we can now see that *Liquid Assets'* predictions have been astonishingly accurate. The '89s turned out to be a truly excellent vintage and the '90s were even better.

How can it be that you'd have two "vintages of the century" in two consecutive years? It turns out that since 1986 there hasn't been a year with below-average growing season temperature. The weather in France has been balmy for more than two decades. A convenient truth for wine lovers is that it's been a great time to grow truly supple Bordeaux.

The traditional experts are now paying a lot more attention to the weather. While many of them have never publicly acknowledged the power of Orley's predictions, their own predictions now correspond much more closely to his simple equation results. Orley still maintains his website, www.liquidasset.com, but he's stopped producing the newsletter. He says, "Unlike the past, the tasters no longer make any horrendous mistakes. Frankly, I kind of killed myself. I don't have as much value added anymore."

Ashenfelter's detractors see him as a heretic. He threatens to de-mystify the world of wine. He eschews flowery and nonsensical ("mus-cular," "tight," or "rakish") jargon, and instead gives you reasons for his predictions.

The industry's intransigence is not just about aesthetics. "The wine dealers and writers just don't want the public informed to the degree that Orley can provide," Kaiser notes. "It started with the '86 vintage. Orley said it was a scam, a dreadful year suffering from a lot of rain and not high enough temperatures. But all the wine writers of the day were raving, insisting it was a great vintage. Orley was right, but it isn't always popular to be right."

Both the wine dealers and writers have a vested interest in main-taining their informational monopoly on the quality of wine. The dealers use the perennially inflated initial rankings as a way to stabilize prices. And *Wine Spectator* and *The Wine Advocate* have millions of dol-lars at stake in remaining the dominant arbiters of quality. As Upton Sinclair (and now Al Gore) has said, "It is difficult to get a man to un-derstand something when his salary depends on his not understanding it." The same holds true for wine. "The livelihood of a lot of people de-pends on wine drinkers believing this equation won't work," Orley says. "They're annoyed because suddenly they're kind of obsolete."

There are some signs of change. Michael Broadbent, chairman of the International Wine Department at Christie's in London, puts it diplomatically: "Many think Orley is a crank, and I suppose he is in many ways. But I have found that, year after year, his ideas and his work are remarkably on the mark. What he does can be really helpful to someone wanting to buy wine."

The Orley Ashenfelter of Baseball

The grand world of oenophiles seems worlds away from the grand-stands of baseball. But in many ways, Ashenfelter was trying to do for wine just what writer Bill James did for baseball.

In his annual *Baseball Abstracts,* James challenged the notion that baseball experts could judge talent simply by watching a player. Michael Lewis's *Moneyball* showed that James was baseball's herald of data-driven decision making. James's simple but powerful thesis was that data-based analysis in baseball was superior to observational expertise:

> The naked eye was inadequate for learning what you need to know to evaluate players. Think about it. One absolutely cannot tell, by watching, the difference between a .300 hitter and a .275 hitter. The difference is one hit every two weeks.... If you see both 15 games a year, there is a 40 percent chance that the .275 hitter will have more hits than the .300 hitter.... The difference between a good hitter and average hitter is simply not visible—it is a matter of record.

Like Ashenfelter, James believed in formulas. He said, "A hitter should be measured by his success in that which he is trying to do, and that which he is trying to do is create runs." So James went out and created a new formula to better measure a hitter's contribution to runs created:

Runs Created = (Hits + Walks) × Total Bases/(At Bats + Walks)

This equation put much more emphasis on a player's on-base per-centage and especially gives higher ratings to those players who tend to walk more often. James's number-crunching approach was particu-lar anathema to scouts. If wine critics like Robert Parker live by their taste and smell, then scouts live and die by their eyes. That's their value added. As described by Lewis:

> In the scouts' view, you found big league ball players by driving sixty thousand miles, staying in a hundred crappy motels, and eating God knows how many meals at Denny's all so you could watch 200 high school and college baseball games inside of four months, 199 of which were completely meaningless to you.... [Y]ou would walk into the ball park, find a seat on the aluminum plank in the fourth row directly behind the catcher and see something no one else had seen—at least no one who knew the meaning of it. You only had to see him once. "If you see it once, it's there."

The scouts and wine critics like Robert Parker have more in common than simply a penchant for spitting. Just as Parker believes that he can assess the quality of a Château's vintage based on a single taste, baseball scouts believe they can assess the quality of the high school prospect based on a single viewing.

In both contexts, people are trying to predict the market value of untested and immature products, whether they be grapes or baseball players. And in both contexts, the central dispute is whether to rely on observational expertise or quantitative data.

Like wine critics, baseball scouts often resort to non-falsifiable euphemisms, such as "He's a real player," or "He's a tools guy."

In *Moneyball,* the conflict between data and traditional expertise came to a head in 2002 when Oakland A's general manager Billy Beane wanted to draft Jeremy Brown. Beane had read James and decided he was going to draft based on hard numbers. Beane loved Jeremy Brown because he had walked more frequently than any other college player. The scouts hated him because he was, well, fat. If he tried to run in corduroys, an A's scout sneered, "he'd start a fire." The scouts thought there was no way that someone with his body could play major league ball. Beane couldn't care less how a player looked. His drafting mantra was "We're not selling jeans." Beane just wanted to win games. The scouts seem to be wrong. Brown has progressed faster than anyone else the A's drafted that year. In September 2006, he was called up for his major league debut with the A's and batted .300 (with an on-base percentage of .364).

There are striking parallels between the ways that Ashenfelter and James originally tried to disseminate their number-crunching results. Just like Ashenfelter, James began by placing small ads for his first newsletter, *Baseball Abstracts* (which he euphemistically characterized as a book). In the first year, he sold a total of seventy-five copies. Just as Ashenfelter was locked out of *Wine Spectator,* James was given the cold shoulder by the *Elias Sports Bureau* when he asked to share data.

But James and Ashenfelter have forever left their marks upon their industries. The perennial success of the Oakland A's, detailed in *Moneyball,* and even the first world championship of the Boston Red Sox under the data-driven management of Theo Epstein are tributes to James's lasting influence. The improved weather-driven predictions of even traditional wine writers are silent tributes to Ashenfelter's impact.

Both have even given rise to gearhead groups that revel in their brand of number crunching. James inspired SABR, the Society for American Baseball Research. Baseball number crunching now even has its own name, sabermetrics. In 2006, Ashenfelter in turn helped launch the *Journal of Wine Economics*. There's even an Association of Wine Economists now. Unsurprisingly, Ashenfelter is its first president. By the way, Ashenfelter's first predictions in hindsight are looking pretty darn good. I looked up recent auction prices for Château Latour and sure enough the '89s were selling for more than twice the price of the '86s, and 1990 bottles were priced even higher. Take that, Robert Parker.

In Vino Veritas

This book's central claim is that the rise of number crunching in wine and baseball are not isolated events. In fact, the wine and baseball examples are microcosms of the larger themes of this book. We are in a historic moment of horse-versus-locomotive competition, where intuitive and experiential expertise is losing out time and time again to number crunching. In the old days, many decisions were simply based on some mixture of experience and intuition. Experts were ordained because of their decades of individual trial-and-error experience. We could trust that they knew the best way to do things, because they'd done it hundreds of times in the past. Experiential experts had survived and thrived. If you wanted to know what to do, you'd ask the gray-hairs.

Now something is changing. Business and government professionals are relying more and more on databases to guide their decisions. The story of hedge funds is really the story of a new breed of number crunchers—call them Super Crunchers—who have analyzed large datasets to discover empirical correlations between seemingly unrelated things. Want to hedge a large purchase of euros? Turns out you should sell a carefully balanced portfolio of twenty-six other stocks and commodities that might include Wal-Mart stock.

What is Super Crunching? It is statistical analysis that impacts real-world decisions. Super Crunching predictions usually bring together some combination of size, speed, and scale. The sizes of the datasets are really big—both in the number of observations and in the number of variables. The speed of the analysis is increasing. We often witness the real-time crunching of numbers as the data come hot off the press. And the scale of the impact is sometimes truly huge. This isn't a bunch of egghead academics cranking out provocative journal articles. Super Crunching is done by or for decision makers who are looking for a better way to do things.

And when I say that Super Crunchers are using large datasets, I mean really large. Increasingly business and government datasets are being measured not in mega- or gigabytes but in tera- and even petabytes (1,000 terabytes). A terabyte is the equivalent of 1,000 gigabytes. The prefix *tera* comes from the Greek word for monster. A terabyte is truly a monstrously large quantity. The entire Library of Congress is about twenty terabytes of text. Part of the point of this book is that we need to start getting used to this prefix. Wal-Mart's data warehouse, for example, stores more than 570 terabytes. Google has about four petabytes of storage which it is constantly crunching. Tera mining is not Buck Rogers's fantasy—it's being done right now.

In field after field, "intuitivists" and traditional experts are battling Super Crunchers. In medicine, a raging controversy over what is called "evidence-based medicine" boils down to a question of whether treatment choice will be based on statistical analysis or not. The intuitivists are not giving up without a fight. They claim that a database can never capture clinical expertise nurtured over a lifetime of experience, that a regression can never be as good as an emergency room nurse with twenty years of experience who can tell whether a kid looks "hinky."

We tend to think that the chess grandmaster Garry Kasparov lost to the Deep Blue computer because of IBM's smarter software. That software is really a gigantic database that ranks the power of different positions. The speed of the computer is important, but in large part it was the computer's ability to access a database of 700,000 grandmaster chess games that was decisive. Kasparov's intuitions lost out to data-based decision making.

Super Crunchers are not just invading and displacing traditional experts; they're changing our lives. They're not just changing the way that decisions are made; they're changing the decisions themselves. Baseball scouts are losing out to gearheads not just because it's a lot cheaper to crunch numbers than to fly scouts out to Palookaville. The scouts are losing because they make poorer predictions. Super Crunchers and experts, of course, don't always disagree. Number crunching sometimes confirms traditional wisdom. The world isn't so perverse that the traditional experts were wrong 100 percent of the time or were even no better than chance. Still, number crunching is leading decision makers to make different and, by and large, better choices.

Statistical analysis in field after field is uncovering hidden relationships among widely disparate kinds of information. If you're a politician and want to know who is most likely to give you a contribution and what form of solicitation is most likely to be successful, you don't need to guess, follow rules of thumb, or trust grizzled traditionalists. Increasingly, it is possible to tease out measurable effects of separate attributes to tell you what kinds of persuasion are likely to work the best. Trolling through databases can reveal underlying causes that traditional experts never even considered.

Data-based decision making is on the rise all around us:

■ Rental car companies and insurers are refusing service to people with poor credit scores because data mining tells them that credit scores correlate with a higher likelihood of having an accident.

■ Nowadays when a flight is canceled, airlines will skip over their frequent fliers and give the next open seat to the mine-identified customer whose continued business is most at risk. Instead of following a first-come, first-serve rule, companies will condition their behavior on literally dozens of consumer-specific factors.

■ The "No Child Left Behind" Act, which requires schools to adopt teaching methods supported by rigorous data analysis, is causing teachers to spend up to 45 percent of class time training kids to pass standardized tests. Super Crunching is even shifting some teachers toward class lessons where every word is scripted and statistically vetted.

Intuitivists beware. This book will detail a dizzying array of Super Crunching stories and introduce you to the people who are making them happen. The number-crunching revolution isn't just about baseball or even sports in general. It is about all the rest of our lives as well. Many times this Super Crunching revolution is a boon to consumers as it helps sellers and governments make better predictions about who needs what. At other times, however, consumers are playing against a statistically stacked deck. Number crunching can put the little guy at a real disadvantage, since sellers can better predict how much they can squeeze out of us.

Steven D. Levitt and Stephen J. Dubner showed in *Freakonomics* dozens of examples of how statistical analysis of databases can reveal the secret levers of causation. Levitt and John Donohue (both my co-authors and friends, about whom you will hear more later) showed that seemingly unrelated events like the abortion rate in 1970 and the crime rate in 1990 have an important connection. Yet *Freakonomics* didn't talk much about the extent to which quantitative analysis is impacting real-world decisions. In contrast, this book is about just that—the impact of number crunching. Decision makers in- and outside of business are using statistical analysis in ways you'd never imagine to drive all kinds of choices.

All of industry, worldwide, is being remade around the database capacities of modern computers. The expectation (and fear) of the 1950s and '60s—in books like Vance Packard's *The Hidden Persuaders*—that sophisticated social engineering, at the behest of big government and big corporations, was about to take over the world has been suddenly resurrected for a new generation. But where we once expected big government to solve all human problems by command and control, we now observe something similar arising in the form of massive data networks.

Why Me?

I'm a number cruncher myself. Even though I teach law at Yale, I learned econometrics while I was studying at MIT for a Ph.D. I've crunched numbers on everything from bail bonds and kidney transplantation to concealed handguns and reckless sex. You might think that your basic Ivy-tower egghead is completely disconnected from real-world decision making (and yes, I am the kind of absentminded professor who was once so engrossed in writing an article on a train that I went to Poughkeepsie instead of New Haven). Still, even data mining by eggheads can sometimes have an impact on the world.

A few years back Steve Levitt and I teamed up to figure out something very practical—the impact of LoJack on auto theft. LoJack is a small radio transmitter that is hidden in one of many possible locations within a car. When the car is reported stolen the police remotely activate the transmitter, and then specially equipped police cars can track the precise location of the stolen vehicle. LoJack is highly effective as a vehicle-recovery device. LoJack Corporation knew this and proudly advertised a 95 percent recovery rate. But Steve and I wanted to test whether LoJack helped reduce auto theft generally. The problem with lots of anti-theft devices is that they might just shift crime around. If you use "the Club" on your car, it probably doesn't stop crime, it just causes the thief to walk down the street and steal another car. The cool thing about LoJack is that it's hidden. In a city covered by LoJack, a thief doesn't know whether a particular car has it or not.

This is just the kind of perversity that Levitt likes to explore. *Freakonomics* reviewers really got it right when they said that Steve looks at things differently. Several years ago, I had an extra ticket and invited Steve to come with me to see Michael Jordan play with the Chicago Bulls. Steve figured he'd enjoy the game more if he was invested in it, but (in sharp contrast to me) he didn't care that much about whether the Bulls won or lost. So just before the game, he hopped online and placed a substantial bet that Chicago would win. Now he really was invested in the game. The online bet changed his incentives.

In an odd way, LoJack is also a device for changing incentives. Before LoJack, many professional thieves were almost untouchable. LoJack changed all that. With LoJack, cops not only recover the vehicle, they often catch the thief. In Los Angeles alone, LoJack has broken up more than 100 chop shops. If you steal 100 cars in a LoJack town, you're almost certain to steal some that have LoJack in them. We wanted to test whether LoJack scared thieves from taking cars generally. If it does, LoJack creates what economists call a "positive externality." When you put the Club on your car, you probably are increasing the chance the next guy's car will be stolen. If enough people put LoJack in their cars, however, Steve and I thought that they might be helping their neighbors by scaring professional car thieves from taking any cars.

Our biggest problem was convincing LoJack to share any of its sales data with us. I remember repeatedly calling and trying to convince them that if Steve and I were right, it would provide another reason for people to buy LoJack. If LoJack reduces the chance that thieves will take *other* people's cars, then LoJack might be able to convince insurance companies to give LoJack users more substantial discounts. A junior executive finally did send us tons of helpful data. But to be honest, LoJack just wasn't that interested in the research at first.

All that changed when they saw the first draft of our paper. After looking at auto theft in fifty-six cities over fourteen years, we found that LoJack had a huge positive benefit for other people. In high-crime areas, a $500 investment in LoJack reduced the car theft losses of non-LoJack users by $5,000. Because we had LoJack sales broken down by both year and city, we could generate a pretty accurate estimate about the proportion of cars with LoJack that were on the road. (For example, in Boston, where the state mandated the largest insurance discount, over 10 percent of the cars had LoJack.) We looked to see what happened to auto theft in the city as a whole as the number of LoJack users increased. Since LoJack service began in different cities in different years, we could estimate the impact of LoJack separate from the general level of crime in that year. In city after city, as the percentage of cars with LoJack increased, the rate of auto theft fell dramatically. Insurance companies weren't giving nearly big enough discounts for LoJack, because they weren't taking into account how much LoJack reduced payouts on even unprotected cars.

Steve and I never bought LoJack stock (because we didn't want to change our own incentives, to tell the truth) but we knew we were sitting on valuable information. When our working paper went public the stock jumped 2.4 percent. Our study has helped convince other cities to adopt the LoJack technology and has spurred slightly higher insurance discounts (but they're still not nearly large enough!).

The bottom line here is that I care passionately about number crunching. I have been a cook myself in the data-mining café. Like Ashenfelter, I am the editor of a serious journal, the *Journal of Law, Economics, and Organization,* where I have to evaluate the quality of statistical papers all the time. I'm well placed to explore the rise of data-based decision-making because I have been both a participant and an observer. I know where the bodies are buried.

Plan of Attack

The next five chapters will detail the rise of Super Crunching across society. The first three chapters will introduce you to two fundamental statistical techniques—regressions and randomized trials—and show how the art of quantitative prediction is reshaping business and government. We'll explore the debate over "evidence-based" medicine in Chapter 4. And Chapter 5 will look at hundreds of tests evaluating how data-based decision making fares in comparison with experience- and intuition-based decisions.

The second part of the book will step back and assess the significance of this trend. We'll explore why it's happening now and whether we should be happy about it. Chapter 7 will look at who's losing out—in terms of both status and discretion. And finally, Chapter 8 will look to the future. The rise of Super Crunching doesn't mean the end of intuition or the unimportance of on-the-job experience. Rather, we are likely to see a new era where the best and the brightest are comfortable with both statistics and ideas.

In the end, this book will not try to bury intuition or experiential expertise as norms of decision making, but will show how intuition and experience are evolving to interact with data-based decision making. In fact, there is a new breed of innovative Super Crunchers—people like Steve Levitt—who toggle between their intuitions and number crunching to see farther than either intuitivists or gearheads ever could before.

CHAPTER 1

Who's Doing Your Thinking for You?

Recommendations make life a lot easier. Want to know what movie to rent? The traditional way was to ask a friend or to see whether reviewers gave it a thumbs-up.

Nowadays people are looking for Internet guidance drawn from the behavior of the masses. Some of these "preference engines" are simple lists of what's most popular. The *New York Times* lists the "most emailed articles." iTunes lists the top downloaded songs. Del.icio.us lists the most popular Internet bookmarks. These simple filters often let surfers zero in on the greatest hits.

Some recommendation software goes a step further and tries to tell you what people like you enjoyed. Amazon.com tells you that people who bought *The Da Vinci Code* also bought *Holy Blood, Holy Grail*. Netflix gives you recommendations that are contingent on the movies that you yourself have recommended in the past. This is truly "collaborative filtering," because your ratings of movies help Netflix make better recommendations to others and their ratings help Netflix make better recommendations to you. The Internet is a perfect vehicle for this service because it's really cheap for an Internet retailer to keep track of customer behavior and to automatically aggregate, analyze, and display this information for subsequent customers.

Of course, these algorithms aren't perfect. A bachelor buying a one-time gift for a baby could, for example, trigger the program into recommending more baby products in the future. Wal-Mart had to apologize when people who searched for *Martin Luther King: I Have a Dream* were told they might also appreciate a *Planet of the Apes* DVD collection. Amazon.com similarly offended some customers who searched for "abortion" and were asked "Did you mean adoption?" The adoption question was generated automatically simply because many past customers who searched for abortion had also searched for adoption.

Still, on net, collaborative filters have been a huge boon for both consumers and retailers. At Netflix, nearly two-thirds of the rented films are recommended by the site. And recommended films are rated half a star higher (on Netflix's five-star ranking system) than films that people rent outside the recommendation system.

While lists of most-emailed articles and best-sellers tend to concentrate usage, the great thing about the more personally tailored recommendations is that they diversify usage. Netflix can recommend different movies to different people. As a result, more than 90 percent of the titles in its 50,000-movie catalog are rented at least monthly. Collaborative filters let sellers access what Chris Anderson calls the "long tail" of the preference distribution. The Netflix recommendations let its customers put themselves in rarefied market niches that used to be hard to find.

The same thing is happening with music. At Pandora.com, users can type in a song or an artist that they like and almost instantaneously the website starts streaming song after song in the same genre. Do you like Cyndi Lauper and Smash Mouth? *Voilà*, Pandora creates a Lauper/Smash Mouth radio station just for you that plays these artists plus others that sound like them. As each song is playing, you have the option of teaching the software more about what you like by clicking "I really like this song" or "Don't play this type of song again."

It's amazing how well this site works for both me and my kids. It not only plays music that each of us enjoys, but it also finds music that we like by groups we've never heard of. For example, because I told Pandora that I like Bruce Springsteen, it created a radio station that started playing the Boss and other well-known artists, but after a few songs it had me grooving to "Now" by Keaton Simons (and because of on-hand quick links, it's easy to buy the song or album on iTunes or Amazon). This is the long tail in action because there's no way a nerd like me would have come across this guy on my own. A similar preference system lets Rhapsody.com play more than 90 percent of its catalog of a million songs every month.

MSNBC.com has recently added its own "recommended stories" feature. It uses a cookie to keep track of the sixteen articles you've most recently read and uses automated text analysis to predict what new stories you'll want to read. It's surprising how accurate a sixteen-story history can be in kickstarting your morning reading. It's also a bit embarrassing: in my case *American Idol* articles are automatically recommended.

Still, Chicago law professor Cass Sunstein worries that there's a social cost to exploiting the long tail. The more successful these personalized filters are, the more we as a citizenry are deprived of a common experience. Nicholas Negroponte, MIT professor and guru of media technology, sees in these "personalized news" features the emergence of the "Daily Me"—news publications that expose citizens only to information that fits with their narrowly preconceived preferences. Of course, self-filtering of the news has been with us for a long time. Vice President Cheney only watches Fox News. Ralph Nader reads *Mother Jones*. The difference is that now technology is creating listener censorship that is diabolically more powerful. Websites like Excite.com and Zatso.net started to allow users to produce "the newspaper of me" and "a personalized newscast." The goal is to create a place "where you decide what's the news." Google News allows you to personalize your newsgroups. Email alerts and RSS feeds allow you to select "This Is the News I Want." If we want, we can now be relieved of the hassle of even glancing at those pesky news articles about social issues that we'd rather ignore.

All of these collaborative filters are examples of what James Surowiecki called "The Wisdom of Crowds." In some contexts, collective predictions are more accurate than the best estimate that any member of the group could achieve. For example, imagine that you offer a $100 prize to a college class for the student with the best estimate of the number of pennies in a jar. The wisdom of the group can be found simply by calculating their average estimate. It's been shown repeatedly that this average estimate is very likely to be closer to the truth than any of the individual estimates. Some people guess too high, and others too low—but collectively the high and low estimates tend to cancel out. Groups can often make better predictions than individuals.

On the TV show *Who Wants to Be a Millionaire,* "asking the audience" produces the right answer more than 90 percent of the time (while phoning an individual friend produces the right answer less than two-thirds of the time). Collaborative filtering is a kind of tailored audience polling. People who are like you can make pretty accurate guesses about what types of music or movies you'll like. Preference databases are powerful ways to improve personal decision making.

eHarmony Sings a New Tune

There is a new wave of prediction that utilizes the wisdom of crowds in a way that goes beyond conscious preferences. The rise of eHarmony is the discovery of a new wisdom of crowds through Super Crunching. Unlike traditional dating services that solicit and match people based on their conscious and articulated preferences, eHarmony tries to find out what kind of person you are and then matches you with others who the data say are most compatible. eHarmony looks at a large database of information to see what types of personalities actually are happy together as couples.

Neil Clark Warren, eHarmony's founder and driving force, studied more than 5,000 married people in the late 1990s. Warren patented a predictive statistical model of compatibility based on twenty-nine different variables related to a person's emotional temperament, social style, cognitive mode, and relationship skills.

eHarmony's approach relies on the mother of Super Crunching techniques—the regression. A regression is a statistical procedure that takes raw historical data and estimates how various causal factors influence a single variable of interest. In eHarmony's case the variable of interest is how compatible a couple is likely to be. And the causal factors are twenty-nine emotional, social, and cognitive attributes of each person in the couple.

The regression technique was developed more than 100 years ago by Francis Galton, a cousin of Charles Darwin. Galton estimated the first regression line way back in 1877. Remember Orley Ashenfelter's simple equation to predict the quality of wine? That equation came from a regression. Galton's very first regression was also agricultural. He estimated a formula to predict the size of sweet pea seeds based on the size of their parent seeds. Galton found that the offspring of large seeds tended to be larger than the offspring of average or small seeds, but they weren't quite as large as their large parents.

Galton calculated a different regression equation and found a similar tendency for the heights of sons and fathers. The sons of tall fathers were taller than average but not quite as tall as their fathers. In terms of the regression equation, this means that the formula predicting a son's height will multiply the father's height by some factor less than one. In fact, Galton estimated that every additional inch that a father was above average only contributed two-thirds of an inch to the son's predicted height.

He found the pattern again when he calculated the regression equation estimating the relationship between the IQ of parents and children. The children of smart parents were smarter than the average person but not as smart as their folks. The very term "regression" doesn't have anything to do with the technique itself. Dalton just called the technique a regression because the first things that he happened to estimate displayed this tendency—what Galton called "regression toward mediocrity"—and what we now call "regression toward the mean."

The regression literally produces an equation that best fits the data. Even though the regression equation is estimated using historical data, the equation can be used to predict what will happen in the future. Dalton's first equation predicted seed and child size as a function of their progenitors' size. Orley Ashenfelter's wine equation predicted how temperature and rain would impact wine quality.

eHarmony produced a formula to predict preference. Unlike the Netflix or Amazon preference engines, the eHarmony regression is trying to match compatible people by using personality and character traits that people may not even know they have or be able to articulate. Indeed, eHarmony might match you with someone who you might never have imagined that you could like. This is the wisdom of crowds that goes beyond the conscious choices of individual members to see what works at unconscious, hidden levels.

eHarmony is not alone in trying to use data-driven matching. Perfectmatch matches users based on a modified version of the Myers-Briggs personality test. In the 1940s, Isabel Briggs Myers and her mother Katharine Briggs developed a test based on psychiatrist Carl Jung's theory of personality types. The Myers-Briggs test classifies people into sixteen different basic types. Perfectmatch uses this M-B classification to pair people who have personalities that historically have the highest probability of forming lasting relationships.

Not to be outdone, True.com collects data from its clients on ninety-nine relationship factors and feeds the results into a regression formula to calculate the compatibility index score between any two members. In essence, True.com will tell you the likelihood you will get along with anyone else.

While all three services crunch numbers to make their compatibility predictions, their results are markedly different. eHarmony believes in finding people who are a lot like you. "What our research kept saying," Warren has observed, "is [to] find somebody whose intelligence is a lot like yours, whose ambition is a lot like yours, whose energy is a lot like yours, whose spirituality is a lot like yours, whose curiosity is a lot like yours. It was a similarity model."

Perfectmatch and True.com in contrast look for complementary personalities. "We all know, not just in our heart of hearts, but in our experience, that sometimes we're attracted [to], indeed get along better with, somebody different from us," says Pepper Schwartz, the empiricist behind Perfectmatch. "So the nice thing about the Myers-Briggs was it's not just characteristics, but how they fit together."

This disagreement over results isn't the way data-driven decision making is supposed to work. The data should be able to adjudicate whether similar or complementary people make better matches. It's hard to tell who's right, because the industry keeps its analysis and the data on which the analysis is based a tightly held secret. Unlike the data from a bunch of my studies (on taxicab tipping, affirmative action, and concealed handguns) that anyone can freely download from the Internet, the data behind the matching rules at the Internet dating services are proprietary.

Mark Thompson, who developed Yahoo! Personals, says it's impractical to apply social science standards to the market. "The peer-review system is not going to apply here," Thompson says. "We had two months to develop the system for Yahoo! We literally worked around the clock. We did studies on 50,000 people."

The matching sites, meanwhile, are starting to compete on validating their claims. True.com emphasizes that it is the only site which had its methodology certified by an independent auditor. True.com's chief psychologist James Houran is particularly dismissive of eHarmony's data claims. "I've seen no evidence they even conducted any study that forms the basis of their test," Houran says. "If you're touting that you're doing something scientific . . . you inform the academic community."

eHarmony is responding by providing some evidence that their matching system works. It sponsored a Harris poll suggesting that eHarmony is now producing about ninety marriages a day (that's over 30,000 a year). This is better than nothing, but it's only a modest success because with more than five million members, these marriages represent about only a 1 percent chance that your $50 fee will produce a walk down the aisle. The competitors are quick to dismiss the marriage number. Yahoo!'s Thompson has said you have a better chance of finding your future spouse if you "go hang out at the Safeway."

eHarmony also claims that it has evidence that its married couples are in fact more compatible. Its researchers presented last year to the American Psychological Society their finding that married couples who found each other through eHarmony were significantly happier than couples married for a similar length of time who met by other means. There are some serious weaknesses with this study, but the big news for me is that the major matching sites are not just Super Crunching to develop their algorithms; they're Super Crunching to prove that their algorithms got it right.

The matching algorithms of these services aren't, however, completely data-driven. All the services rely at least partially on the conscious preferences of their clients (regardless of whether these preferences are valid predictors of compatibility). eHarmony allows clients to discriminate on the race of potential mates. Even though it's only acting on the wishes of its clients, matching services that discriminate by race may violate a statute dating back to the Civil War that prohibits race discrimination in contracting. Think about it. eHarmony is a for-profit company that takes $50 from black clients and refuses to treat them the same (match them with the same people) as some white clients. A restaurant would be in a lot of trouble if it refused to seat Hispanic customers in a section where customers had stated a preference to have "Anglos only."

eHarmony has gotten into even more trouble for its refusal to match same-sex couples. The founder's wife and senior vice president, Marilyn Warren, has claimed that "eHarmony is meant for everybody. We do not discriminate in any way." This is clearly false. They would refuse to match two men even if, based on their answers to the company's 436 questions, the computer algorithm picked them to be the most compatible. There's a sad irony here. eHarmony, unlike its competitors, insists that similar people are the best matches. When it comes to gender, it insists that opposites attract. Out of the top ten matching sites, eHarmony is the only one that doesn't offer same-sex matching.

Why is eHarmony so out of step? Its refusal to match gay and lesbian clients, even in Massachusetts where same-sex marriage is legal, seems counter to the company's professed goal of helping people find lasting and satisfying marriage partners. Warren is a self-described "passionate Christian" who for years worked closely with James Dobson's Focus on the Family. eHarmony is only willing to facilitate certain types of legal marriages regardless of what the statistical algorithm says. In fact, because the algorithm is not public, it is possible that eHarmony puts a normative finger on the scale to favor certain clients.

But the big idea behind these new matching services—the insight they all share—is that data-based decision making doesn't need to be limited to the conscious preferences of the masses. Instead, it is possible to study the results of decisions and tease out from inside the data the factors that lead to success. This chapter is about how simple regressions are changing decisions by improving predictions. By sifting through aggregations of data, the regression technique can uncover the levers of causation that are hidden to casual and even expert observation. And even when experts feel that a particular factor is an important determinant of some outcome, the regression technique literally can price it out.

Just for fun, Garth Sundem, in his book *Geek Logik,* used a regression to create a formula to predict how long celebrity marriages will last. (It turns out that having more Google hits reduces a marriage's chances—especially if the top Google hits include sexually suggestive photos!) eHarmony, Perfectmatch, and True.com are doing the same kind of thing, but they're doing it for profit. These services are engaged in a new kind of Super Crunching competition. The game's afoot and it's a very different kind of game.

Harrah's Feels Your Pain

The same kind of statistical matchmaking is also happening inside companies like Lowe's and Circuit City, which are using Super Crunching to select job applicants. Employers want to predict which job applicants are going to make a commitment to their job. Unlike traditional aptitude tests that try to suss out an applicant's IQ, the modern tests are much closer to eHarmony's questionnaire in trying to evaluate three underlying personality traits of the applicants: their conscientiousness, agreeableness, and extroversion. Data mining shows that these personality traits are better predictors of worker productivity (especially turnover) than more traditional ability testing. Barbara Ehrenreich was appalled when she took an employment test at a Minneapolis Wal-Mart and was told that she had given the wrong answer when she agreed with the proposition "there is room in every corporation for a non-conformist." Yet regressions suggest that people who think Wal-Mart is for non-conformists aren't a good fit and are more likely to turn over. It's one thing to argue that Wal-Mart and other employers should reorganize their mind-numbing jobs to make them less boring. But in a world where mind-numbing jobs are legal, it's hard for me to see what's wrong with a statistically validated test that helps match employees that are most compatible with those jobs.

Mining for non-obvious predictors is not just about hiring good applicants. It's also helping businesses keep their costs down, especially the costs of stagnant inventory. Businesses that can do a better job of predicting demand can do a better job of predicting when they are about to run out of something. And it can be just as important for businesses to know when they're *not* about to run out of something. Instead of bearing the costs of large inventories lying around, Super Crunching allows firms to move to just-in-time purchasing. Stores like Wal-Mart and Target try to get as close as possible to having no excess inventory on hand at all. "What they have on the shelf is what they've got," said Scott Gnau, general manager of the data-mining company Teradata. "If I buy six cans of yellow corn off the shelf, and there are now three cans left, somebody knows that happened immediately so they can make sure that the truck coming my way gets some more corn loaded on it. It's gotten to the point that as you're putting stuff in your trunk, the store is loading the truck at the distribution center." These prediction strategies can be based on highly specific details about likely demand. Before Hurricane Ivan hit Florida in 2004, Wal-Mart already had started rushing strawberry Pop-Tarts to stores in the hurricane's path. Analyzing sales of other stores in areas hit by hurricanes, Wal-Mart was able to predict that people would be yearning for the gooey comfort of Pop-Tarts, finger food that doesn't require cooking or refrigeration. Firms are engaging in "analytic competition" in an explicit attempt to out-data-mine the other guy, struggling to first uncover and then exploit the hidden determinants of profitability.

Some of this Super Crunching is done in-house, but truly large datasets are warehoused and analyzed by specialist firms like Teradata, which manages literally terabytes of data. Sixty-five percent of the top worldwide retailers (including Wal-Mart and JCPenney) use Teradata. More than 70 percent of airlines and 40 percent of banks are its customers.

Crunching terabytes helps predict which customers are likely to defect to rivals. For its most profitable customers, Continental Airlines keeps track of every negative experience that may increase the chance of defection. The next time a customer who experienced a bad flight takes to the air, a data-mining program automatically kicks in and gives the crew a heads-up. Kelly Cook, Continental's onetime director of customer relationship management, explains, "Recently, a flight attendant walked up to a customer flying from Dallas to Houston and said, 'What would you like to drink? And, oh, by the way, I am so sorry we lost your bag yesterday coming from Chicago.' The customer flipped."

UPS uses a more sophisticated algorithm to predict when a customer is likely to switch to another shipping company. The same kind of regression formula that we saw at play with wines and matchmaking is used to predict when a customer's loyalty is at risk, and UPS kicks into action even before the customer has thought about switching. A salesperson proactively calls the customer to reestablish the relationship and resolve potential problems, dramatically reducing the loss of accounts.

Harrah's casinos are particularly sophisticated at predicting how much money they can extract from clients and still retain their business. Harrah's "Total Rewards" customers use a swipeable electronic card that lets Harrah's capture information on every game played at every Harrah's casino they've visited. Harrah's knows in real time on a hand-by-hand (or slot-by-slot) basis how much each player is winning or losing. It combines these gambling data together with other information such as the customer's age and the average income in the area where he or she lives, all in a data warehouse.

Harrah's uses this information to predict how much a particular gambler can lose and still enjoy the experience enough to come back for more. It calls this magic number the "pain point." And once again, the pain point is calculated by plugging customer attributes into a regression formula. Given that Shelly, who likes to play the slots, is a thirty-four-year-old white female from an upper-middle-class neighborhood, the system might predict her pain point for an evening of gambling is a $900 loss. As she gambles, if the database senses that Shelly is approaching $900 in slot losses, a "luck ambassador" is dispatched to pull her away from the machine.

"You come in, swipe your card, and are sitting at a slot," Teradata's Gnau said. "When you get close to that pain point, they come out and say, 'I see you're having a rough day. I know you like our steakhouse. Here, I'd like you to take your wife to dinner on us right now.' So it's no longer pain. It becomes a good experience."

To some, this kind of manipulation is the science of diabolically separating as many dollars from a customer as possible on a repeated basis. To others, it is the science of improving customer satisfaction and loyalty—and of making sure the right customers get rewarded. It's actually a bit of both. I'm troubled that Harrah's is making what can be an addictive and ruinous experience even more pleasurable. But because of Harrah's pain-point predictions, its customers tend to leave happier.

The Harrah's strategy of targeting benefits is being adopted in different retail markets. Teradata found, for example, that one of its airline clients was giving perks to its frequent fliers based solely on how many miles they flew each year, with Platinum customers getting the most benefits. But the airline hadn't taken account of how profitable these customers were. They didn't plug in other available information, such as how much Platinum fliers paid for tickets, where they bought them, whether they called customer service, and most important, whether they traveled on flights where the airline actually made money. After Teradata crunched the numbers taking into account these bottom-line attributes, the airline found out that almost all of its Platinum fliers were unprofitable. Teradata's Scott Gnau summed it up, "So they were giving people an incentive to make them lose money."

The advent of tera mining means that the era of the free lunch is over. Instead of having more profitable customers subsidizing the less profitable, firms will be able to target rewards to their most profitable customers. But *caveat emptor*! In this brave new world, you should be scared when a firm like Harrah's or Continental becomes particularly solicitous of your business. It probably means you have been paying too much. Airlines are learning to give upgrades and other favorable treatment to the customers that make them the most money, not just the ones that fly the most. Airlines can then "encourage people to become more profitable," Gnau explains, by charging you more, for example, for buying tickets through a call center than for buying them online.

This hyper-individualized segmentation of consumers also lets firms offer new personalized services that clearly benefit society. Progressive insurance capitalizes on the new capabilities of data mining to define extremely narrow groups of customers, e.g., motorcycle riders ages thirty and above, with college educations, credit scores over a certain level, and no accidents. For each cell, the company runs regressions to identify factors that most closely correlate with that group's losses. Super Crunching on this radically expanded set of factors lets it set prices on types of consumers who were traditionally written off as uninsurable.

Super Crunching has also created a new science of extraction. Data mining increases firms' ability to charge individualized prices that predict our individualized pain points. If your walk-away price is higher than mine, tera mining will lead firms to take a bigger chunk out of you one way or another. In a Super Crunching world, consumers can't afford to be asleep at the wheel. It's no longer safe to rely on the fact that other consumers care about price. Firms are figuring out more and more sophisticated ways to treat the price-oblivious differently than the price-conscious.

Tell Me What You Know About Me

Tera mining sometimes gives businesses a decided information advantage over their customers. Hertz, after analyzing terabytes of sales data, knows a lot more than you do about how much gas you're likely to leave in the tank if you prepay for the gas. Cingular knows the probability that you will go beyond your "anytime minutes" or leave some unused. Best Buy knows the probability that you will make a claim on an extended warranty. Blockbuster knows the probability that you will return the rental late.

In each of these cases, the companies not only know the general-ized probability of some behavior, they can make incredibly accurate predictions about how individual customers are going to behave. The power of corporate tera mining creepily suggests the opening lines of Psalm 139:

> *You have searched me and you know me.*
> *You know when I sit and when I rise; you perceive my thoughts from afar.*
> *You discern my going out and my lying down; you are familiar with all*
>   *my ways.*

We may have free will, but data mining can let business emulate a kind of aggregate omniscience. Indeed, because of Super Crunching, firms sometimes may be able to make more accurate predictions about how you'll behave than you could ever make yourself.

But instead of trying to prohibit statistical analysis, we might re-act to the possibility of advantage-taking by simply making sure that consumers know that the number crunching is going on. The rise of these predictive models suggests the possibility of a new kind of dis-closure duty. Usually government only requires firms to tell a con-sumer about their products or services ("made in Japan"). Now firms sometimes know more about consumers than the consumers know about themselves. We could require firms to educate consumers about themselves. It might be helpful if Avis told you, before you agree to prepay for gasoline, that other people like you tend to leave more than a third of a tank full when they return the car—so that the effective price for prepaid gas is really four bucks per gallon. Or Verizon might be asked to tell you when their statistical model predicts that you're on the wrong phone plan.

Government could also Super Crunch some of its enormous datasets to inform citizens about themselves. Indeed, Super Crunching could truly help reinvent government. The IRS nowadays is almost universally disliked. Yet the IRS has tons of information that could help people if only it would analyze and disseminate the results.

Imagine a world where people looked to the IRS as a source for useful information. The IRS could tell a small business that it might be spending too much on advertising or tell an individual that the average taxpayer in her income bracket gave more to charity or made a larger IRA contribution. Heck, the IRS could probably produce fairly accurate estimates about the probability that small businesses (or even marriages) would fail. In fact, I'm told that Visa already does predict the probability of divorce based on credit card purchases (so that it can make better predictions of default risk). Of course, this is all a bit Orwellian. I might not particularly want to get a note from the IRS saying my marriage is at risk. (A little later on, we will take on whether all this Super Crunching is really a good idea. Just because it's possible to make accurate predictions about intimate matters doesn't mean that we should.) But I might at least want the option of having the government make predictions about various aspects of my life. Instead of thinking of the IRS as solely a taker, we might also think of it as an information provider. We could even change its name to the "Information & Revenue Service."

Consumers Fight Back

Even without government's help, entrepreneurs are bringing new services to market which use Super Crunching as a consumer advocacy tool. Coming to the aid of consumers, these firms are using data-crunching to counteract the excesses of seller-side price extraction. The airline industry is especially fertile ground for such advocacy, because airlines engage in increasingly bewildering pricing shenanigans—trying to find in their databases any crevice of an opportunity to enhance their "revenue yield."

What's a consumer to do? Enter Oren Etzioni, a professor of computer science at the University of Washington. On a fateful day in 2002, Etzioni was peeved to learn that the people sitting next to him on an airplane trip had bought their tickets for a much lower price merely because they waited to buy their tickets later. He had a student go out and try to forecast whether particular airline fares would increase or decrease as you got closer to the travel date. With just a little data, the student could make pretty accurate forecasts about whether it was a good idea to buy early or wait.

Etzioni ran with the idea in a big way. What he did is a prime example of how consumer-oriented Super Crunching can counteract the number-crunching price manipulations of sellers. He created Farecast.com, a travel website that lets you search for the lowest current fare. Farecast goes further than other fare-search sites; it adds an arrow that simply points up or down telling you which way Farecast predicts fares are headed. Even a prediction that the fare is likely to go up is valuable, because it lets consumers know that they should hurry up and pull the trigger.

"We're doing the same thing the weatherman does," said Hugh Crean, Farecast's chief executive. "We haven't achieved clairvoyance, nor will we. But we're doing travel search with a real level of advocacy for the consumer." Henry H. Harteveldt, a vice president and principal travel analyst at Forrester Research in Cambridge, says Farecast is trying to level the informational playing field for travelers. "Farecast provides guidance, much like a stockbroker, about whether you should take action now, or whether you should wait."

The company (which was originally named Hamlet and had the motto "to buy or not to buy") is based on a serious Super Crunch. In a five-terabyte database, it keeps fifty billion prices that it purchased from ITA Software, a company that sells price data to travel agents, websites, and computer reservation services. Farecast has information on nearly all the major carriers except Jet Blue and Southwest (who do not provide data to ITA). Farecast can indirectly account for and even predict Jet Blue and Southwest pricing by looking at how other airlines on the same routes react to price changes of the two missing competitors.

Farecast bases its predictions on 115 indicators that are reweighed every day for every market. It pays attention not just to historic pricing patterns, but also to a host of factors that will shift the demand or supply of tickets—things like the price of fuel or the weather or who wins the National League pennant. It turns all this information into an up-arrow if it predicts the price will go up, or a down-arrow if it predicts the price will go down. "It's like going to the ballet," Harteveldt says. "We don't see the many years of practice and toil and blood and sweat and strain that the ballet dancer has experienced. We're only there in the auditorium watching them as they dance gracefully on the stage. With Farecast, we see the graceful dancing onstage. We don't see the data-crunching, we don't really care about the data-crunching."

Farecast turns the tera-crunching tables on the airlines. It uses the same databases and even some of the same statistical techniques that airlines have been using to extract money from consumers. But Farecast isn't the only service that has been crunching numbers to help the little guy.

There are a bunch of other services popping up that crunch large datasets to predict prices. Zillow.com in just a few months has become one of the most visited real estate sites on the net. Zillow crunches a dataset of over sixty-seven million home prices to help both buyers and sellers price their homes.

And if you can predict the selling price of a house, why not the selling price of a PDA? Accenture is doing just that. Rayid Ghani, a researcher at Accenture's Information Technology group, for the past two years has been mining the data from 50,000 eBay auctions to predict the price that PalmPilots and other PDAs will ultimately sell for. He hopes to convince insurance companies or eBay itself to offer sellers price-protection insurance that guarantees a minimum price they'll receive. Explains Ghani, "You'll put a nice item on eBay. Then if you pay me ten dollars, I'll guarantee it will go for at least a thousand dollars. And if it doesn't, I'll pay you the difference." Of course, auction bidders will also be interested in these predictions. Bidcast software that will suggest whether you should bid now or wait for the next item is sure to be coming to a web portal near you.

Sometimes Super Crunching is helping consumers just get through the day. Inrix's "Dust Network" crunches data on the speed of a half million commercial vehicles to predict traffic jams. Today's large commercial fleets of taxis and delivery vans are equipped with global positioning systems that in real time can relay information not just about their position but about how fast they're going. Inrix combines this traffic-flow information with information about the weather, accidents, and even when schools and rock concerts are letting out, to provide instantaneous advice on the fastest way to get from point A to point B.

Meanwhile, Ghani is working to use Super Crunching to personalize our shopping experience further. Soon, supermarkets may ask us to swipe our loyalty cards as we enter the store—at which point the store will data mine through our previous shopping trips and make a prediction of what foods we're running out of. Ghani sees a day when the supermarket will become a food shopping advisor, telling us what we need to buy and offering special deals for the day's shopping trip.

The simple predictive power of a good data crunch can be applied to almost any activity where people do the same thing again and again. Super Crunching can be used to give one side an edge in a commercial transaction, but there's no reason why it has to be the seller. As more and more data becomes increasingly available for free, consumer services like Farecast and Zillow will step forward and crunch it.

In Regressions We Trust

These services not only tell you which way the price is going to move, they also tell you how confident they are in their estimates. So with Farecast a consumer might learn not only that the fare is expected to drop, but also that this type of prediction turns out to be correct

80 percent of the time. Farecast knows that it doesn't always have enough data to make a very precise prediction. Other times it does. So it lets you know not only its best guess, but how confident it is in that guess. Farecast not only tells you how confident it is, but it puts its money where its mouth is. For $10, it will provide you with "Fareguard" insurance—which guarantees that an offered airfare will remain valid for a week, or Farecast will make up the difference.

This ability to report a confidence level in predictions underscores one of the most amazing things about the regression technique. The statistical regression not only produces a prediction, it also simultaneously reports how precisely it was able to predict. That's right—a regression tells you how accurate the prediction is. Sometimes there are just not enough historical data to make a very precise estimate and the output of the regression technique tells you just this. Indeed, it gets even better, because the regression tells you not only the precision of the regression equation on the whole, it also tells you the precision with which it was able to estimate the impact of each individual term in the regression equation.

So Wal-Mart learns three different kinds of things from its employment test regression. First, it learns how long a particular applicant is likely to stay on the job. Second, it learns how precisely it made this prediction. The predicted longevity of an applicant might be thirty months, but the regression will separately report the probability that the applicant would work less than fifteen months. If the thirty months prediction is fairly accurate, the probability that the applicant will work only fifteen months would be pretty small, but for inaccurate predictions this probability might begin to balloon. A lot of people want to know whether they can really trust a regression prediction. If the prediction is imprecise (say because of poor or incomplete data), the regression itself will be the first one to tell you not to rely on it. When was the last time you heard a traditional expert tell you the precision of his or her estimate?

And finally, the regression output tells Wal-Mart how precisely it was able to measure the impact of individual parts of the regression equation. Wal-Mart isn't about to report the results of its regression formula. However, the regression output might tell Wal-Mart that applicants who think "there is room in every corporation for a non-conformist" are likely to work 2.8 months less than people who disagree. The prediction associated with the specific question is 2.8 fewer months, holding everything else about the applicant constant. The regression output can go even further and tell Wal-Mart the chance that "non-conformist" applicants will end up working *longer*. Depending on the accuracy of the 2.8-month prediction, this probability or a contrary influence might be 2 percent or 40 percent. The regression begins the process of validating itself. It tells you the impact of more rainfall on wine, and whether that particular influence is really valid.

All the World's a Mine

Tera mining of customer records, airline prices, and inventories is peanuts compared to Google's goal of organizing all the world's information. Google reportedly has five petabytes of storage capacity. That's a whopping 5,000 terabytes (or a quadrillion bytes). At first, it may not seem that a search engine really has much to do with data mining. Google makes a concordance of all the words used on the Internet and then if you search for "kumquat," it simply sends you a list of all the web pages that use that word the most times. Yet Google uses all kinds of Super Crunching to help you find the kumquat pages you really want to see.

Google has developed a Personalized Search feature that uses your past search history to further refine what you really have in mind. If Bill Gates and Martha Stewart both Google "blackberry," Gates is more likely to see web pages about the email device at the top of his results list, while Stewart is more likely to see web pages about the fruit.

Google is pushing this personalized data mining into almost every one of its features. Its new web accelerator dramatically speeds up access to the Internet—not by some breakthrough in hardware or software technology—but by predicting what you are going to want to read next. Google's web accelerator is continually pre-picking web pages from the net. So while you're reading the first page of an article, it's already downloading pages two and three. And even before you fire up your browser tomorrow morning, simple data mining helps Google predict what sites you're going to want to look at (hint: it's probably the same sites that you look at most days).

Yahoo! and Microsoft are desperately trying to play catch-up in this analytic competition. Google has deservedly become a verb. I'm frankly in awe of how it has improved my life. Nonetheless, we Internet users are fickle friends. The search engine that can best guess what we're really looking for is likely to win the lion's share of our traffic. If Microsoft or Yahoo! can figure out how to outcrunch Google, they will very quickly take its place. To the Super Crunching victor go the web traffic spoils.

Guilt by Association

The granddaddy of all of Google's Super Crunching is its vaunted PageRank. Among all the web pages that include the word "kumquat," Google will rank a page higher if it has more web pages that are linking to it. To Google, every link to a page is a kind of vote for that web page. And not all votes are equal. Votes cast by web pages that are themselves important are weighted more heavily than links from web pages that have low PageRanks (because no one else links to them).

Google found that web pages with higher PageRanks were more likely to contain the information that users are actually seeking. And it's very hard for users to manipulate their own PageRank. Merely creating a bunch of new web pages that link to your home page won't work because only links from web pages that themselves have reasonably high PageRanks will have an impact. And it's not so easy to create web pages that other sites will actually link to.

The PageRank system is a form of what webheads call "social network analysis." It's a good kind of guilt by association. Social network analysis can also be used as a forensic tool by law enforcement to help identify actual bad guys.

I've used this kind of data mining myself.

A couple of years ago, my cell phone was stolen. I hopped on the Internet and downloaded the record of telephone calls that were made both to and from my phone. This is where network analysis came into play. The thief made more than a hundred calls before my service was cut off. Yet most of the calls were to or from just a few phone numbers. The thief made more than thirty calls to one phone number, and that phone number had called into the phone several times as well. When I called that number, a voice mailbox told me that I'd reached Jessica's cell phone. The third most frequent number connected me with Jessica's mother (who was rather distraught to learn that her daughter had been calling a stolen phone).

Not all the numbers were helpful. The thief had called a local weather recording a bunch of times. By the fifth call, however, I found someone who said he'd help me get my phone back. And he did. A few hours later, he handed it back to me at a McDonald's parking lot. Just knowing the telephone numbers that a bad guy calls can help you figure out who the bad guy is. In fact, cell phone records were used in just this way to finger the two men who killed Michael Jordan's father.

This kind of network analysis is also behind one of our nation's efforts to smoke out terrorists. *USA Today* reported that the National Security Agency has been amassing a database with the records of two trillion telephone calls since 2001. We're talking thousands of terabytes of information. By finding out who "people of interest" are calling, the NSA may be able to identify the players in a terrorist network and the structure of the network itself.

Just like I used the pattern of phone records to identify the bad guy who stole my phone, Valdis Krebs used network analysis of public information to show that all nineteen of the 9/11 hijackers were within two email or phone call connections to two al-Qaeda members whom the CIA already knew about before the attack. Of course, it's a lot easier to see patterns after the fact, but just knowing a probable bad guy may be enough to put statistical investigators on the right track.

The 64,000-terabyte question is whether it's possible to start with just a single suspect and credibly identify a prospective conspiracy based on an analysis of social network patterns. The Pentagon is understandably not telling whether its data-mining contractors—which include our friend Teradata—have succeeded or not. Still, my own experience as a forensic economist working to smoke out criminal fraud makes me more sanguine that Super Crunching may prospectively contribute to homeland security.

Looking for Magic Numbers

A few years ago, Peter Pope, who was then the inspector general of the New York City School Construction Authority, called me and asked for help. The Construction Authority was spending about a billion dollars a year in a ten-year plan to renovate New York City schools. Many of the schools were in terrible disrepair and a lot of the money was being used on "envelope" work—roof and exterior repairs to maintain the integrity of the shell of the building. New York City had a long and sordid history of construction corruption and bid rigging, so the New York state legislature had created a new position of inspector general to put an end to inflated costs and waste.

Peter was a recent law grad who was interested in doing a very different kind of public interest law. Making sure that construction auctions and contract change-orders are on the up-and-up is not as glamorous as taking on a death penalty case or making a Supreme Court oral argument, but Peter was trying to make sure that thousands of schoolchildren had a decent place to go to school. He and his staff were literally risking their lives. Organized crime is not happy when someone comes in and rocks their boat. Once Peter was on the scene, nothing was business as usual.

Peter called me because he had discovered a specific type of fraud that had been taking place in some of his construction auctions. He called it the "magic number" scam.

During the summer of 1992, Elias Meris, the principal owner of the Meris Construction Corporation, was under investigation by the Internal Revenue Service. Meris agreed, in exchange for IRS leniency, to wear a wire and provide information on a bid-rigging scam involving School Construction Authority employees and other contractors. Working undercover for prosecutors, Meris taped conversations with senior project officer John Dransfield and a contract specialist named Mark Parker.

The contract specialist is the person who publicly opens the sealed bids of contractors one at a time at a procurement auction and reads out loud the price that a contractor has bid.

In the "magic number" scam, the bribing bidder would submit a sealed bid with the absolute lowest price at which it would be willing to do the project. At the public bid openings, Parker would save the dishonest contractor's bid for last and, knowing the current low bid, he would read aloud a false bid just below this price, so that the briber would win but would be paid just slightly less than the bidder who honestly should have won. Then Dransfield would use Wite-Out to doctor the briber's bid—writing in the amount that Parker had read out loud. (If the lowest real bid turned out to be below the lowest amount at which the dishonest bidder wanted the job, the contract specialist wouldn't use the Wite-Out and would just read the dishonest bidder's written bid.) This "magic number" scam allowed dishonest bidders to win the contract whenever they were willing to do the job for less than the lowest true bid, but they would be paid the highest possible price.

Pope's investigation eventually implicated in the scam eleven individuals within seven contracting firms. Next time you're considering renovating your New York pied-à-terre, you might want to avoid

Christ Gatzonis Electrical Contractor Inc., GTS Contracting Corp., Batex Contracting Corp., American Construction Management Corp., Wolff & Munier Inc., Simins Falotico Group, and CZK Construction Corp. These seven firms used the "magic number" scam to win at least fifty-three construction auctions with winning bids totaling over twenty-three million dollars.

Pope found these bad guys, but he called me to see if statistical analysis could point the finger toward other examples of "magic number" fraud. Together with auction guru Peter Cramton and a talented young graduate student named Alan Ingraham, we ran regressions to see if particular contract specialists were cheating.

This is really looking for needles in a haystack. It is doubtful that a specialist would rig all of his auctions. The key for us was to look for auctions where the difference between the lowest and second-lowest bid was unusually small. Using statistical regressions that controlled for a host of other variables—including the number of bidders and an engineer's pre-auction estimate of cost as well as the third-lowest bid placed in the auction—Alan Ingraham identified two new contract specialists who presided over auctions where there was a disturbingly small difference between the winning and the second-lowest bid. Without knowing even the names of the contract specialists (the inspector general's data referred to them by number only), we were able to point the inspector general's office in a new direction. Alan turned the work into two chapters of his doctoral dissertation. While the results of the inspector general's investigation are confidential, Peter was deeply appreciative and earlier this year thanked me for "helping us catch two more crooks."

This "magic number" story shows how Super Crunching can reveal the past. Super Crunching also can predict what you will want and what you will do. The stories of eHarmony and Harrah's, magic numbers, and Farecast are all stories of how regressions have slipped the bounds of academia and are being used to predict all kinds of things.

The regression formula is "plug and play"—plug in the specified attributes and, *voilà,* out pops your prediction. Of course, not all predictions are equally valuable. A river can't rise above its source and regression predictions can't overcome insufficient data. If your dataset is too small, no regression in the world is going to make very accurate predictions. Still, unlike intuitivists, regressions know their own limitations and can answer Ed Koch's old campaign question, "How Am I Doing?"

CHAPTER 2

Creating Your Own Data
with the Flip of a Coin

In 1925, Ronald Fisher, the father of modern
statistics, formally proposed using random assignments to test whether particular medical interventions had some predicted effect. The first
randomized trial on humans (of an early antibi-
otic against tuberculosis) didn't take place until
the late 1940s. But now, with the encouragement
of the Food and Drug Administration, randomized tests have become the gold standard for
proving whether or not medical treatments are
efficacious.

This chapter is about how business is playing
catch-up. Smart businesses know that regression
equations can help them make better predictions.
But for the first time, we're also starting to see
businesses combine regression predictions with
predictions based on their own randomized trials.
Businesses are starting to go out and create their
own data by flipping coins. We'll see that random-

ized testing is becoming an important tool for data-driven decision making. Like the new regression studies, it's Super Crunching to answer the bottom-line questions of what works. The poster child for the power of combining these two core Super Crunching tools is a company that made famous the question "What's in Your Wallet?"

Capital One, one of the nation's largest issuers of credit cards, has been at the forefront of the Super Crunching revolution. More than 2.5 million people call CapOne each month. And they're ready for your call.

When you call CapOne, a recording immediately prompts you to enter your card number. Even before the service representative's phone rings, a computer algorithm kicks in and analyzes dozens of characteristics about the account and about you, the account holder. Super Crunching sometimes lets them answer your question even before you ask it.

CapOne found that some customers call each month just to find out their balance or to see whether their payment has arrived. The computer keeps track of who makes these calls, and routes them to an automated system that answers the phone this way: "The amount now due on your account is $164.27. If you have a billing question, press 1...." Or: "Your last payment was received on February 9. If you need to speak with a customer-service representative, press 1...." A phone call that might have taken twenty or thirty seconds, or even a minute, now lasts just ten seconds. Everyone wins.

Super Crunching also has transformed customer service calls into a sales opportunity. Data analysis of customer characteristics generates a list of products and services that this kind of consumer is most willing to buy, and the service rep sees the list as soon as she takes the call. It's just like Amazon's "customers who like this, also like this" feature, but transmitted through the rep. Capital One now makes more than a million sales a year through customer-service marketing—and their data-mining predictions are the big reason why. Again, everybody wins.

But maybe not equally. CapOne gives itself the lion's share of the gains whenever possible. For example, a statistically validated algorithm kicks in whenever a customer tries to cancel her account. If the customer is not so valuable, she is routed to an automated service where she can press a few buttons and cancel. If the customer has been (or is predicted to be) profitable, the computer routes her to a "retention specialist" and generates a list of sweeteners that can be offered.

When Nancy from North Carolina called to close her account because she felt her 16.9 percent interest rate was too high, CapOne routed her call to a retention specialist named Tim Gorman. CapOne's computer automatically showed Tim a range of three lower interest rates—ranging from 9.9 percent to 12.9 percent—that he could offer to keep Nancy's business.

When Nancy claimed on the phone that she just got a card with a 9.9 percent rate, Tim responded with "Well, ma'am, I could lower your rate to 12.9 percent." Because of Super Crunching, CapOne knows that a lot of people will be satisfied with this reduction (even when they say they've been offered a lower rate from another card). And when Nancy accepts the offer, Tim gets an immediate bonus. Everyone wins. But because of data mining, CapOne wins a bit more.

CapOne Rolls the Dice

What really sets CapOne apart is its willingness to literally experiment. Instead of being satisfied with a historical analysis of consumer behavior, CapOne proactively intervenes in the market by running randomized experiments.

In 2006, it ran more than 28,000 experiments—28,000 tests of new products, new advertising approaches, and new contract terms.

Is it more effective to print on the outside envelope "LIMITED TIME OFFER" or "2.9 Percent Introductory Rate!"? CapOne answers this question by randomly dividing prospects into two groups and seeing which approach has the highest success rate.

It seems too simple. Yet having a computer flip a coin and treating prospects who come up heads differently than the ones who come up tails is the core idea behind one of the most powerful Super Crunching techniques ever devised.

When you rely on historical data, it is much harder to tease out causation. A miner of historical data who wants to find out whether chemotherapy worked better than radiation would need to control for everything else, such as patient attributes, environmental factors— really anything that might affect the outcome. In a large random study, however, you don't need these controls. Instead of controlling for whether the patients smoked or had a prior stroke, we can trust that in a large randomized division, about the same proportion of smokers will show up in each treatment type.

The sample size is the key. If we get a large enough sample, we can be pretty sure that the group coming up heads will be statistically identical to the group coming up tails. If we then intervene to "treat" the heads differently, we can measure the pure effect of the intervention. Super Crunchers call this the "treatment effect." It's the causal holy grail of number crunching: after randomization makes the two groups identical on every other dimension, we can be confident that any change in the two groups' outcome was *caused* by their different treatment.

CapOne has been running randomized tests for a long time. Way back in 1995, it ran an even larger experiment by generating a mailing list of 600,000 prospects. It randomly divided this pool of people into groups of 100,000 and sent each group one of six different offers that varied the size and duration of the teaser rate. Randomization let CapOne create two types of data. Initially the computerized coin flip was itself a type of data that CapOne created and then relied upon to decide whether to assign a prospect to a particular group. More importantly, the response of these groups was new data that only existed because the experiment artificially perturbed the status quo. Comparing the average response rate of these statistically similar groups let CapOne see the impact of making different offers. Because of this massive study, CapOne learned that offering a teaser rate of 4.9 percent for six months was much more profitable than offering a 7.9 percent rate for twelve months.

Academics have been running randomized experiments inside and outside of medicine for years. But the big change is that businesses are relying on them to reshape corporate policy. They can see what works best and immediately change their corporate strategy. When an academic publishes a paper showing that there's point shaving in basketball, nothing much changes. Yet when a business invests tens of thousands of dollars on a randomized test, they're doing it because they're willing to be guided by the results.

Other companies are starting to get in on the act. In South Africa, Credit Indemnity is one of the largest micro-lenders, with over 150 branches throughout the country. In 2004, it used randomized trials to help market its "cash loans." Like payday loans in the U.S., cash loans are short-term, high-interest credit for the "working poor." These loans are big business in South Africa, where at any time as many as 6.6 million people borrow. The typical loan is only R1000 ($150), about a third of the borrower's monthly income.

Credit Indemnity sent out more than 50,000 direct-mail solicitations to former customers. Like CapOne's mailings, these solicitations offered random interest rates that varied from 3.25 percent to 11.75 percent. As an economist, it was comforting to learn from Credit Indemnity's experiment that yes, there was larger demand for lower priced loans.

Still, price wasn't everything. What was really interesting about the test is that Credit Indemnity simultaneously randomized other aspects of the solicitations. The bank learned that simply adding a photo of a smiling woman in the corner of the solicitation letter raised the response rate of male customers by as much as dropping the interest rate 4.5 percentage points. They found an even bigger effect when they had a marketing research firm call the client a week before the solicitation and simply ask questions: "Would you mind telling us if you anticipate making large purchases in the next few months, things like home repairs, school fees, appliances, ceremonies (weddings, etc.), or even paying off expensive debt?"

Talk about your power of suggestion. Priming people with a

pleasant picture or bringing to mind their possible need for a loan in a non-marketing context dramatically increased their likelihood of responding to the solicitation.

How do we know that the picture or the phone call really caused the higher response rate? Again, the answer is coin flipping. Randomizing over 50,000 people makes sure that, on average, those shown pictures and those not shown pictures were going to be pretty much the same on every other dimension. So any differences in the average response rate between the two groups must be caused by the difference in their treatment.

Of course, randomization doesn't mean that those who were sent photos are each exactly the same as those who were not sent photos. If we looked at the heights of people who received photo solicitations, we would see a bell curve of heights. The point is that we would see the same bell curve of heights for those who received photos without solicitations. Since the *distribution* of both groups becomes increasingly identical as the sample size increases, then we can attribute any differences in the *average* group response to the difference in treatment.

In lab experiments, researchers create data by carefully controlling for everything to create matched pairs that are identical except for the thing being tested. Outside of the lab, it's sometimes simply impossible to create pairs that are the same on all the peripheral dimensions. Randomization is how businesses can create data without creating perfectly matched individual pairs. The process of randomization instead creates matched distributions. Randomization thus allows Super Crunchers to run the equivalent of a controlled test without actually having to laboriously match up and control for dozens or hundreds of potentially confounding variables.

The implications of the randomized marketing trials for lending profitability are pretty obvious. Instead of dropping the interest rate five percentage points, why not simply include a picture? When Credit Indemnity learned the results of the study, they were about to start doing just that. But shortly after the tests were analyzed, the bank was taken over. The new bank not only shut down future testing, it also laid off tons of the Credit Indemnity employees—including those who had been the strongest proponents of testing. Ironically, some of these former employees have taken the lessons of the testing to heart and are now implementing the results in their new jobs working for Credit Indemnity's competitors.

You May Be Watching a Random Web Page

Testing with chance isn't limited to banks and credit companies; indeed, Offermatica.com has turned Internet randomization into a true art form. Two brothers, Matt and James Roche, started Offermatica in 2003 to capitalize on the ease of randomizing on the Internet. Matt is its CEO and James works as the company's president. As their name suggests, Offermatica has automated the testing of offers. Want to know whether one web page design works better than another? Offermatica will set up software so that when people click on your site, either one page or the other will be sent at random. The software then can tell you, in real time, which page gets more "click throughs" and which generates more purchases.

What's more, they can let you conduct multiple tests at once. Just as Credit Indemnity randomly selected the interest rate and independently decided whether or not to include a photo, Offermatica can randomize over multiple dimensions of a web page's design.

For example, Monster.com wanted to test seven different elements of the employers' home pages. They wanted to know things like whether a link should say "Search and Buy Resumes" or just "Search Resumes" or whether there should be a "Learn More" link or not. All in all, Monster had 128 different page permutations that it wanted to test. But by using the "Taguchi Method," Offermatica was able to test just eight "recipe" pages and still make accurate predictions about how the untested 120 other web pages would fare.*

---

* The method is named after Genichi Taguchi, who devised it more than fifty years ago to help manufacturers test multiple aspects of a manufacturing process using just a subset of the tests that had been traditionally required.

Offermatica software not only automates the randomization, it also automatically analyzes the Internet response. In real time, as the test was being conducted, Monster could see a continuously updating graph of not only how the eight recipe pages were faring, but also how untested pages would likely fare in generating actual sales. The lines of each alternative page stretch across the graph like a horse race with clear winners and clear losers. Imagine it—instantaneous information on 128 different treatments with tens of thousands of observations. This is randomization on steroids. Offermatica shows the way that Super Crunching often exploits technology to shorten the time between data collection, analysis, and implementation. With Offermatica, the time between the test and the marketing change can be just a matter of hours.

By the way, if you think you have a good graphic eye, try to see which of these two boxes you think tested better:

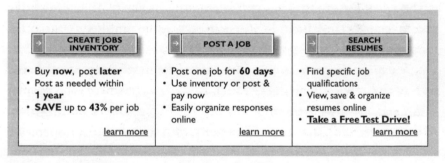

or

SOURCE: Monster.com Scores Millions, http://www.offermatica.com/stories-1.7.html.

I personally find the curved icons of the lower box to be more appealing. That's what Monster thought too. The lower box is the box that Monster actually started with before the test. Yet it turns out that employers spent 8.31 percent more per visit when they were shown the top box. This translates into tens of millions of dollars per year for Monster's ecommerce channel. Instead of trusting its initial instinct, Monster went out, perturbed the status quo, and watched what happened. It created a new type of data and put its initial instinct to the test.

Jo-Ann Fabrics got an even bigger surprise. Part of the power of testing multiple combinations is that it lets companies be bolder, to take bigger risks with test marketing. You might not think that JoAnn.com would draw enough web traffic to make Internet testing feasible, but they pull over a million unique visitors a month. They have enough traffic to do all kinds of testing.

So when JoAnn.com was optimizing their website, they decided to take a gamble and include in their testing an unlikely promotion for sewing machines: "Buy two machines and save 10 percent." They didn't expect this test to pan out. After all, how many people need to buy two sewing machines? Much to their amazement, the promotion generated by far the highest returns. "People were pulling their friends together," says Linsly Donnelly, JoAnn.com's chief operating officer. The discount was turning their customers into sales agents. Overall, randomized testing increased its revenue per visitor by a whopping 209 percent.

In the brick-and-mortar world, the cost of running randomized experiments with large enough samples to give you a statistically significant result sometimes severely limits the number of experiments that can be done. But the Internet changes all this. "As the cost of showing a group of people a given experience gets close to zero," Matt Roche, Offermatica's CEO, says, "the number of experiences you can give is close to infinity."

Seeing how consumers respond to a whole bunch of different online experiences is what Offermatica is all about. It's a wildly different model for deciding who decides. Matt comes alive when he talks about seeing the battle for corporate control of your eyeballs: "I go to meetings where you have all these people sitting around a table claiming authority. You have the analytic guy who has the amulet of historical data. You've got the branding guy who has this mystical certainty about what makes brands stronger. And of course you got the authority of title itself, the boss who's used to thinking that he knows best. But what's missing is the consumer's voice. All these forms of authority are substituting for the customer's voice. What we're really doing at Offermatica is listening to what consumers want."

Offermatica not only has to do battle with the in-house analytic guy who crunches numbers on historical data, it also has to take on "usability experts" who run hyper-controlled experiments in university laboratories. The usability experts are sure of certain axioms that have been established in the lab—things like "people look at the upper left-hand corner first" or "people look at red more than blue." Roche responds, "In the real world, an ad is competing against so many other inputs. There's no such thing as a controlled experiment. They cling to a sandcastle of truth in a tsunami of other information." It's so cheap to test and retest alternatives that there's no good reason to blindly accept the wisdom of academic axioms.

It shouldn't surprise you that the smarts at Google are also riding the randomization express. Like Offermatica, they make it easy to give consumers different ad experiences and then see which ads they like best. Want to know whether your AdWords ad for beer should say "Tastes Great" or "Less Filling"? Well, Google will put both ads in rotation and then tell you which ad people are more likely to click through. Since the order in which people run Google searches is fairly random, alternating the order of two ads will have the same effect as randomizing. Indeed, Google even will start by rotating your ads and then automatically shift toward the ad that has the higher click-through rate.

I just ran this test to help decide what to name this book. *The End of Intuition* was the original working title for this book, but I wondered whether *Super Crunchers* might instead better convey the book's positive message. So I set up a Google AdWords campaign. Anyone searching for words like "data mining" or "number crunching" would be shown either

Super Crunchers		**The End of Intuition**
Why Thinking-by-Numbers	**OR**	Why Thinking-by-Numbers
Is the New Way to Be Smart		Is the New Way to Be Smart
www.bantamdell.com		www.bantamdell.com

I found that random viewers were 63 percent more likely to click through on the *Super Crunchers* ad. (They also substantially preferred the current subtitle to "Why Data-Driven Decision Making Is the New Way to Be Smart.") In just a few days, we had real-world reactions from more than a quarter of a million page views. That was good enough for me. I'm proud to be able to say that *Super Crunchers* is itself a product of Super Crunching.

Who Is Usefully Creative?

A common feature of all the foregoing random trials is that someone still has to come up with the alternatives to be tested. Someone has to have the idea to try to sell two sewing machines or the idea to have a research firm call a week in advance. The random trial method is not the end of intuition. Instead it puts intuition to the test.

In the old days, firms would have to bet the ranch on a national television campaign. On the web, you can roll the dice on a number of different campaigns and quickly shift to the campaign that produces the best results. The creative process is still important, but creativity is literally an input to the testing process.

In fact, the AdWords randomization feature could provide a great test of who can write the most effective ad. Ad agencies might think of testing applicants by seeing who is able to improve on a client's

Google ad. Imagine an episode of *The Apprentice* where the contestants were ranked on their objective ability to optimize the mass market sales of some popular web page.

The potential for randomized web testing is almost limitless. Randomized trials of alternatives have increased not just click-through rates and sales, they've increased the rate at which web forms are completed. Randomization can be used to enhance the performance of any web page.

That includes the layout of online newspapers. The graphic designers at Slate, MSNBC, even the *New York Times* could learn a thing or two from randomized testing. In fact James Roche, the president of Offermatica, says that they've started to do some work for web publications. They are typically brought in by the subscription department. However, once "the editors see the increase in online subscriptions," James explains, they "warm to the idea of using Offermatica to optimize their primary business drivers: page views and ad clicks."

Charities and political campaigns could also test which web designs optimize their contributions. In fact, charities have already started using off-line randomized trials to explore the wellsprings of giving. Experimental economists Dean Karlan and John List helped a non-profit advocacy group test the effectiveness of direct mailings much the same way as Credit Indemnity did. They sent out over 50,000 letters to past contributors asking for gifts. The letters differed in whether and how they offered a matching gift. Some of the letters offered no matching gift, some offered dollar-for-dollar matching, and some letters offered two- and even three-for-one matching. Dollar-for-dollar matching offers did increase giving by about 19 percent. The surprising result, however, was that the two-for-one and three-for-one matches didn't generate any higher giving than the one-for-one match. This simple study gives charities a powerful new tool. A donor who wants the biggest bang for the buck would do a lot better to offer it as part of a one-for-one program.

One thing we've seen over and over is that decision makers overestimate the power of their own intuitions. The intuitions make sense to us and we become wedded to them. Randomized testing is an objective way to see whether we were right. And testing is a road that never ends. Tastes change. What worked yesterday may not work tomorrow. A system of periodic retesting with randomized trials is a way to ensure that your marketing efforts remain optimized. Super Crunching is centrally about data-driven decisions. And ongoing randomized trials make sure that there is a continual supply of new data to drive decisions.

Randomization—It's Not Just for Breakfast

You might think at this point that the power of randomization is just about marketing—optimizing direct-mail solicitations or web advertisements. But you'd be wrong. Randomized trials are also being used to help manage both employee and customer relationships.

Kelly Cook, director of customer relationship management at Continental Airlines, used the coin-flipping approach to figure out how to build stronger customer loyalty. She wanted to see how best to respond when a passenger experienced what Continental euphemistically called a "transportation event." This is the kind of event you don't want to experience, such as having your flight severely delayed or canceled.

Cook randomly assigned Continental customers who had endured transportation events to one of three groups. For the next eight months, one group received a form letter apologizing for the event. The second group received the letter of apology and compensation in the form of a trial membership in Continental's President's Club. And the third group, which served as a control, received nothing.

When the groups were asked about their experience with Continental, the control group that didn't receive anything was still pretty angry. "But the other groups' reaction was amazement that a company would have written them unsolicited to say they were sorry,"

Cook recalls. The two groups that received a letter spent 8 percent more on Continental tickets in the ensuing year. For just the 4,000 customers receiving letters, that translated to extra revenue of $6 million. Since expanding this program to the top 10 percent of Continental's customers, the airline has seen $150 million in additional revenues from customers who otherwise would have had a good reason to look elsewhere.

Just sending a letter without compensation was enough to change consumer perceptions and behavior. And the compensation of trial membership itself turned into a new source of profit. Thirty percent of customers who received a trial membership in Continental's President's Club decided to renew their membership after the trial period expired.

Yet retailers beware. Customers can become angry if they learn that the same product is being offered at different prices. In September 2000, the press started running with the story of a guy who said that when he deleted the cookies on his computer (which identified him as a regular Amazon customer), Amazon's quoted price for DVDs fell from $26.24 to $22.74. A lot of customers were suddenly worried that Amazon was rigging the Internet. The company quickly apologized, saying that the difference was the result of a random price test. In no uncertain terms CEO Jeff Bezos declared, "We've never tested and we never will test prices based on customer demographics."

He also announced a new policy with regard to random testing that might serve as a model for other companies that test. "If we ever do such a test again, we'll automatically give customers who purchase a test item the lowest test price for that item at the conclusion of the test period—thereby ensuring that all customers pay the lowest price available." Just because you test higher prices doesn't mean that you have to actually charge them once a customer places an order.

These randomized tests about prices are a lot more problematic than other kinds of randomization. When Offermatica randomizes over different visual elements of a web page, it's usually about trying to improve the customer experience by removing unnecessary obstacles. The results of this kind of experiment are truly a win–win for both the seller and the buyers. But as we saw with CapOne, randomization can also be used to test how much money can be extracted from the consumer. Offermatica or AdWords could be used to run randomized tests about what kind of price the market will bear.

More insidiously, sellers might run randomized tests on what kinds of terms to include in their consumer contracts. Some businesses are really pushing the ethical and legal envelope by testing whether to conceal warranty waivers on their website. If a company runs a randomized test and finds that disclosing the waiver on a top-level page substantially reduces their sales, they have an incentive to bury the offensive information. Matt Roche emphasizes, though, that even this kind of warranty testing is a two-edged sword that sometimes helps consumers. Randomized tests show that displaying certain types of express warranties increases sales. "Including the *Verisign* security certificate almost universally gives you a lift," he said. "Tests show that consumers demand trust. Without this hard evidence, a lot of companies wouldn't have been willing to give it."

Getting Into the Game

Randomized trials are not some ivory tower daydream. Lots of firms are already using them. The puzzle is why more firms aren't following the lead of CapOne and Jo-Ann Fabrics. Why isn't Wal-Mart running randomized experiments? Wal-Mart is great at using what their consumers have done in the past to predict the future. Yet at least publicly they won't admit to randomized testing. All too often information management is limited to historical data, to recent and not-so-recent information about past transactions. Business is now very good at tracking these kinds of data, but businesses as a group still haven't gone far enough in proactively creating useful new data.

The examples in this chapter show that it's really not that hard. The Excel "=rand()" function on just about any computer will flip the coin for you. Any bright high schooler could run a randomized test and analyze it. The setup isn't hard and the analysis is just a comparison of two averages, the average results for the "treated" and "untreated" groups. That's really all Offermatica is doing when they tell you the average click-through rate for one web page versus another. (Okay, it's more complicated when they use the Taguchi Method to analyze the results of multiple tests.)

The simplicity of these studies in both collecting the data and analyzing the results is also why it's so much easier to explain the results to people who don't like thinking about things like heteroskedasticity and BLUE estimators. The fancier statistical regressions are much harder for non-statisticians to understand and trust. In effect, the statistician at some point has to say, "Trust me. I did the multivariate regression correctly." It's easier to trust a randomized experiment. The audience still has to trust that the researcher flipped the coin correctly, but that's basically it. If the only difference between the two groups is how they're treated, then it's pretty clear that the treatment is the only thing that can be causing a difference in outcome.

Randomization also frees the researcher to take control of the questions being asked and to create the information that she wants. Data mining on the historic record is limited by what people have actually done. Historic data can't tell you whether teaching statistics in junior high school increases math scores if no junior high has ever offered this subject. Super Crunchers who run randomized experiments, however, can create information to answer this question by randomly assigning some students to take the class (and seeing if they do better than those who were not chosen).

Firms historically have been willing to create more qualitative data. Focus groups are one way to explore what the average "man on the street" thinks about a new or old product. But the marketer of the future will adopt not just social science methods of multivariate regressions and the mining of historical databases, she will start exploiting the randomized trials of science.

Businesses realize that information has value. Your databases not only help you make better decisions, database information is a commodity that can be sold to others. So it's natural that firms are keeping better track of what they and their customers are doing. But firms should more proactively figure out what pieces of information are missing and take action to fill the gaps in their data. And there's no better way to create information about what causes what than via randomized testing.

Why aren't more firms doing this already? Of course, it may be because traditional experts are just defending their turf. They don't want to have to put their pet policies to a definitive test, because they don't want to take the chance that they might fail. But in part, the relative foot-dragging may have to do with timing. Randomized trials require firms to hypothesize in advance before the test starts. My editor and I had a devil of a time deciding what book titles we really wanted to test. Running regressions, in contrast, lets the researcher sit back and decide what to test after the fact. Randomizers need to take more initiative than people who run after-the-fact regressions, and this difference might explain the slower diffusion of randomized trials in corporate America.

As we move to a world in which quantitative information is increasingly a commodity to be hoarded, or bought, or sold, we will increasingly find firms that start to run randomized trials on advertisements, on price, on product attributes, on employment policies. Of course not all decisions can be pre-tested. Some decisions are all-or-nothing, like the first moon launch, or whether to invest $100 million in a new technology. Still, for many, many different decisions, powerful new information on the wellsprings of human action is just waiting to be created.

This book is about the leakage of social science methods from the academy to the world of on-the-ground decision making. Usually and unsurprisingly, business has been way ahead of government in picking up useful technologies. And the same is true about the technology of Super Crunching. When there's a buck to be made, businesses more than bureaucrats scoop it up. But randomization is one place where government has taken the lead. Somewhat perversely, the checks and balances of a two-party system might have given government a leg up on the more unified control of corporations in the move to embrace randomization. Political adversaries who can't agree on substance can at least reach bipartisan agreement on a procedure for randomization. They'll let some states randomly test their opponents' preferred policy if the opponents will let other states randomly test their preferred policy. Bureaucrats who lack the votes to get their favored policy approved sometimes have sufficient support to fund a randomized demonstration project. These demonstration projects usually start small, but the results of randomized policy trials can have supersized impacts on subsequent policy.

CHAPTER 3

Government by Chance

Way back in 1966, Heather Ross, an economics graduate student at MIT, had an audacious idea. She applied for a huge government grant to run a randomized test of the Negative Income Tax (NIT). The NIT pays you money if your income falls below some minimum level and effectively guarantees people a minimum income regardless of how much they earn working. Heather wanted to see whether the NIT reduced people's incentives to work. When the Office of Economic Opportunity approved her grant, Heather ended up with a $5 million thesis. She found that the NIT didn't reduce employment nearly as much as people feared, but there was a very unexpected spike in divorce. Poor families that were randomly selected to receive the NIT were more likely to split up.

The biggest impact of Ross's test was on the process of how to evaluate government programs themselves. Heather's simple application of the medical randomization method to a policy issue unleashed what has now grown into a torrent of hundreds of randomized public policy experiments at home and abroad. U.S. lawmakers increasingly have come to embrace randomization as the best way to test what works. Acceptance of the randomized methodology was not a partisan issue but a neutral principle to separate the good from the bad and the ugly. Government isn't just paying for randomized trials; for the first time the results are starting to drive public policy.

Spending Money to Save Money

In 1993, a young whiz-kid economist named Larry Katz had a problem. As chief economist for the Department of Labor, he was trying to convince Congress that it could save $2 billion a year by making a simple change to unemployment insurance (UI). By spending some additional money to give the unemployed job-search assistance, Larry thought we could reduce the length of time that workers made unemployment claims. The idea that spending money on a new training program would save $2 billion was not an easy sale to skeptical politicians.

Larry, however, is no pushover. He's not physically intimidating. Even now, in his forties, Larry still looks more like a wiry teenager than a chaired Harvard professor (which he actually is). But he is scary smart. Long ago, Larry and I were roommates at MIT. I still remember how, in our first week of grad school, the TA asked an impossible question that required us to use something called the hypergeometric distribution—something we'd never studied. To most of us, the problem was impenetrable. Katz, however, somehow derived the distribution on his own and solved the problem.

While Larry is soft-spoken and calm, he is absolutely tenacious when he is defending an idea that he knows is right. And Larry knew he was right about job assistance because of randomized testing. His

secret weapon was a series of welfare-to-work tests that states conducted after Senator Patrick Moynihan in 1989 inserted an evidence-
based provision into our federal code. The provision said that a state
could experiment with new ideas on how to reduce unemployment insurance if and only if the ideas were supported by an evaluation plan
that must "include groups of project participants and control groups
assigned at random in the field trial."

Moynihan's mischief led to more than a dozen randomized demonstration projects. Many of the states ran tests looking to see whether
providing job-search assistance could reduce a state's unemployment
insurance payments. Instead of providing job training for new on-the-
job skills, the job-search assistance was geared to provide advice on
how to go about applying and interviewing for a new job. These
"search-assistance" tests (which took place in Minnesota, Nevada,
New Jersey, South Carolina, and Washington) were also novel because
they combined the two central types of database–decision making, regressions and randomization.

The search-assistance programs used regression equations to predict
which workers were most likely to have trouble finding a job on their
own. The regressions represent a kind of statistical profiling which allowed programs to concentrate their efforts to help those workers who
were likely to need it and respond to it most. After the profiling stage,
randomization kicked in. The tests randomly assigned qualifying unemployed workers to treatment and control groups to allow a direct test of
the specific intervention's impact. So the studies used both regressions
and randomization to promote better public policy.

These UI tests taught us a lot about what works and what doesn't.
Unemployed workers who received the assistance found a new job
about a week earlier than similar individuals who did not receive the
assistance. In Minnesota, which provided the most intensive search
assistance, the program reduced unemployment by a whopping four
weeks. Finding jobs faster didn't mean finding poorer-paying jobs
either. The jobs found by the program participants paid just as well as
the ones found later by non-participants.

Most importantly from the government's perspective, the programs more than paid for themselves. The reduction in UI benefits paid plus the increase in tax receipts from faster reemployment were more than enough to pay for the cost of providing the search assistance. For every dollar invested in job assistance, the government saved about two dollars.

Larry used the results of this testing to convince Congress that the savings from mandated job assistance would be more than enough to fund the projected $2 billion needed to extend unemployment benefits during the recession of 1993. Larry calmly deflated any objection Congressional leadership raised about the program's effectiveness. The transparency of randomized trials made Larry's job a lot easier. Unemployed workers who were similar in every respect found jobs faster when they received search assistance. It's pretty manifest that search assistance was the reason why. Up against the brutal power of a randomized trial and the intellect of Katz, the opponents never really had a chance.

The statistical targeting of these search-assistance programs was crucial to their cost savings. The costs of government support programs can balloon if they are opened to too large a class of individuals. This is the same kind of problem Head Start has to grapple with. Lots of people will tell you that prison costs three times as much as early childhood education. The problem with this comparison is that there are a lot more kids than there are prisoners. It's not necessarily cheaper to pay for every kid's pre-K program even if pre-K programs reduce the chance of later criminality—because most kids aren't going to commit crime anyway. Racial profiling by police has gotten a deservedly bad name, but non-racial profiling may be the key to concentrating resources on the kids who are high risk. You might think that you can't say which kids at four or five are at higher risk of committing crime when they're sixteen or seventeen, but isn't that just what Orley Ashenfelter did with regard to immature grapes? The search-assistance tests show that statistical profiling may be used for smarter targeting of government support.

A True State Laboratory of Ideas

The growth of randomized experiments like the search-assistance trials for the first time pays off on the idea that our federalist system of independent states could create a rigorous "laboratory of democracy." The laboratory metaphor is that each state could experiment with what it thought were the best laws and collectively we could sit back and learn from each other. Great idea. The problem is the experimental design. A good experiment requires a good control group. The problem with many state experiments is that we don't know how exactly to compare the results. Alaska's not really like Arizona. The states offer a great opportunity for experimentation. But in real labs, you don't let the rats design the experiment. The move to randomization lets states experiment on themselves, but with a procedure that provides the necessary control group. Now states are creating quality data that can be used to implement data-driven policy making. Unlike earlier case studies that are quietly shelved and forgotten, the randomized policy experiments are more likely to have Super Crunching impacts on real-world decisions.

And the pace of randomized testing has accelerated. Hundreds of policy experiments are now under way. Larry is one of the leaders of a new HUD-funded effort to find out what happens if poor families are given housing vouchers that can only be used in low-poverty (middle-class) neighborhoods. This "Move to Opportunity" (MTO) test randomly gave housing vouchers to very low-income families in five cities (Baltimore, Boston, Chicago, Los Angeles, and New York City) and is collecting information for ten years on how vouchers impact everything from employment and school success to health and crime. The results aren't all in yet, but the first returns suggest that there is no huge educational or crime-reduction benefit from moving poor kids to more affluent neighborhoods (with more affluent schools). Girls who moved were a little more successful in school and healthier, but boys who moved have done worse in school and are more likely to commit crime. Regardless of where the cards ultimately fall, the MTO data are going to provide policy makers for the first time with very basic information about whether changing your neighborhood can change your life.

Randomized trials are instructing politicians not just on what kinds of policies to enact, but on how to get themselves elected in the first place. Political scientists Donald Green and Alan Gerber have been bringing randomized testing to bear on the science of politics. Want to know how best to get out the vote? Run a randomized field experiment. Want to know whether direct mail or telephone solicitations work best? Run a test. Want to know how negative radio ads will influence turnout of both your and your opponent's voters? Run the ads at random in some cities and not in others.

Keeping an Eye Out for Chance

The sky is really the limit. Any policy that can be applied at random to some people and not others is susceptible to randomized tests. Randomized testing doesn't work well for the Federal Reserve's interest rate setting—because it's hard to subject some of us to high and others to low interest rates. And it won't help us design the space shuttle. We're not going to send some shuttles up with plastic O-rings and others with metal O-rings. But there are tons of business and government policies that are readily susceptible to random assignment.

Up to now, I've been describing how business and government analysts have intentionally used randomized assignments to test for impacts. However, it's also possible for Super Crunchers to piggyback on randomized processes that were instituted for other purposes. There are in fact over 3,000 state statutes that already explicitly mandate random procedures. Instead of flipping coins to create data, we can sometimes look at the effects of randomized processes that were independently created. Because some colleges randomly assigned roommates, it became possible to test the impact roommates have on one another's drinking. Because California randomizes the order that candidates appear on the ballot, it became possible to test the impact of having your name appear at the top (it turns out that appearing first helps a lot in primaries, not so much in general elections where people are more apt to vote the party line).

But by far the most powerful use of pre-existing randomization concerns the random assignment of judges to criminal trials. For years it has been standard procedure in federal courts to randomly assign cases to the sitting trial judges in that jurisdiction. As with the alphabet lottery, random case assignment was instituted as a way of assuring fairness (and deterring corruption).

In the hands of Joel Waldfogel, randomization in criminal trials has become a tool for answering one of the most central questions of criminal law—do longer sentences increase or decrease the chance that a prisoner will commit another crime?

Waldfogel is an auburn-haired, slightly balding imp who has a reputation for being one of the funniest number crunchers around. And he has one of the quirkiest minds. Joel often shines his light on overlooked corners of our society. Waldfogel has looked at how game show contestants learned from one season to the next. And he has estimated "the deadweight loss" of Christmas—that's when your aunt spends a lot on a sweater that you just can't stand. He's the kind of guy who goes out and ranks business schools based on their value added— how much different schools increase the expected salary of their respective students.

To my mind, his most important number crunching has been to look at the sentencing proclivity of judges. Just as we've seen time and time before, random assignment means that each judge within a particular district should expect to see the same kind of cases. Judges in Kansas may see different cases than judges in D.C., but the random assignment of cases assures that judges within any particular district will see not only about the same proportion of civil and criminal cases, they will see about the same proportion of criminal cases where the defendant deserves a really long sentence.

Waldfogel's "a-ha" moment was simply that random judicial assignment would allow him to rank judges on their sentencing proclivity. If judges within a district were seeing the same kinds of cases, then intra-district disparities in criminal sentencing had to be attributable to differences in judicial temperament. Of course, it might be that certain judges by chance just received a bunch of rotten apple defendants who really deserved to go away for a long time. But statistics is really good at distinguishing between noise and an underlying tendency.

Even though federal judges are required to follow sentencing guidelines—grids that foreordain narrow ranges of sentences for defendants who committed certain crimes—Waldfogel found substantial sentencing disparities between judges. There really are the modern-day equivalent of "hanging judges" and "bleeding hearts"—who found ways to manipulate the guidelines to increase or decrease the time served.

These differences in sentencing are troubling if we want our country to provide "equal protection under the law." But Waldfogel and others saw that these disparities at least have one advantage—they give us a powerful way to measure whether longer sentences increased or decreased recidivism.

The holy grail of criminologists has been to learn whether prison "hardens" or "rehabilitates" criminals. Does putting rapists away for ten instead of five years increase or decrease the chance that they'll rape again when they're back out on the street? This is an incredibly hard question to answer because the people we put away for ten years are different from those we put away for five years. Ten-year inmates might have a higher recidivism rate—not because prison hardened them, but because they were worse guys to begin with.

Waldfogel's randomization insight provided a way around this problem. Why not look at the recidivism rates of criminals sentenced by individual judges? Since the judges see the same types of criminals, differences in the judges' recidivism rates must be attributable to disparities in the judges' sentencing. Random assignment to (severe or lenient) judges is equivalent to randomly assigning criminals to longer or shorter sentences. Just as Waldfogel ranked business schools based on how well their students performed in the aftermarket, Waldfogel's exploitation of randomization allows a ranking of judges based on how well their defendants perform in the post-prison aftermarket.

So what's the answer? Well, the best evidence is that neither side in the debate is right. Putting people in jail neither increases nor decreases the probability that they'll commit a crime when they're released. Brookings Institute economist Jeff Kling found that post-release earnings of people sentenced by the hanging judges were not statistically different from those sentenced by the judicial bleeding hearts. A convict's earnings after prison are a pretty strong indicator of recidivism because people who are caught and put back in prison have zero taxable earnings. More recently, two political scientists, Danton Berube and Donald Green, have directly looked at the recidivism rates of those sentenced by judges with different sentencing propensities. Not only do they find that longer sentences incapacitate prisoners from committing crimes outside of prison, but also that the longer sentences of the hanging judges were not associated with increased or decreased recidivism rates once the prisoners hit the streets. The "lock 'em up" crowd can take solace in the fact that longer sentences are not hardening prisoners. Then again, the longer sentences don't specifically deter future bad acts. Because of randomized assignments, we might start changing the debate about sentencing length from questions about specific deterrence and rehabilitation, and instead ask whether longer sentences deter other people from committing crime or whether simply incapacitating bad guys makes longer sentences worthwhile.

But the big takeaway here concerns the possibility of piggybacking. Instead of randomly intervening to create data, it is sometimes possible to piggyback on pre-existing randomization. That's what criminologists are doing with regard to random judicial assignments. And it's what I've started to do with regard to random assignments at our local school district. About 20 percent of New Haven schoolchildren apply to attend oversubscribed magnet schools. For the schools that are oversubscribed, the kids are chosen by lottery. Can you see what piggybacking will let me do? I can look at all the kids who applied to Amistad Academy and then compare the average test scores of those who got in and those who didn't. Piggybacking on randomization provides me with the kind of Super Crunching information that will allow me to rank the value added of just about every school in the district.

The World of Chance

The randomized testing of social policy is truly now a global phenomenon. Dozens upon dozens of regulatory tests have been completed in every corner of the globe. Indeed, if anything, developing countries have taken the lead in embracing the randomizing method. A test that would cost millions of dollars in the U.S. can be undertaken for a fraction of the cost in the Third World.

The spread of randomized testing is also due to the hard work of the Poverty Action Lab. Founded at MIT in 2003 by Abhijit Banerjee, Esther Duflo, and Sendhil Mullainathan, the Poverty Action Lab is devoted to using randomized trials to test what development strategies actually work. Their motto is "translating research into action." By partnering with non-profit organizations around the world, in just a short time the lab has been able to crank out dozens of tests on everything from public health measures and micro-credit lending to AIDS prevention and fertilizer use.

The driving force behind the lab is Esther Duflo. Esther has endless energy. A wiry mountain climber (good enough to summit at Mount Kenya), she also has been rated the best young economist from France, and is the recipient of one of the lucrative MacArthur "Genius" fellowships. Esther has been tireless in convincing NGOs (non-government organizations) to condition their subsidies on randomized testing.

The use of randomized tests to reduce poverty sometimes raises ethical concerns—because some destitute families are capriciously denied the benefits of the treatment. In fact, what could be more capricious than a coin flip? Duflo counters, "In most instances we do not know whether the program will work or whether it's the best use of the money." By conducting a small randomized pilot study, the NGO can figure out whether it's worthwhile to "scale" the project—that is, to apply the intervention on a non-randomized basis to the entire country. Michael Kremer, a lab affiliate, sums it up nicely: "Development goes through a lot of fads. We need to have evidence on what works."

Other countries can sometimes test policies that U.S. courts would never allow. Since 1998, India has mandated that the chief (the *Pradhan*) in one-third of village councils has to be a woman. The villages with set aside (or "reserved") female chiefs were selected at random. *Voilà,* we have another natural experiment which can be analyzed simply by comparing villages with and without chiefs who were mandated to be female. Turns out the mandated women leaders were more likely to invest in infrastructure that was tied to the daily burdens of women—obtaining water and fuel—while male chiefs were more likely to invest in education.

Esther has also helped tackle the problem of rampant teacher absenteeism in Indian schools. A non-profit organization, *Seva Mandir,* has been instrumental in establishing single-teacher schoolhouses in remote rural areas where students have no access to government education. Yet massive teacher absenteeism has undermined the effectiveness of these schools. In some states, teachers simply don't show up for class about half the time.

Esther decided to see if cameras might help. She took 120 of *Seva Mandir*'s single-teacher schools and in half of them she gave the teacher a camera with a tamper-proof date and time stamp. The teachers at the "camera schools" were instructed to have one of the children photograph the teacher with other students at the beginning and end of each school day. The salary of a teacher at the camera school was a direct function of his or her attendance.

A little monitoring went a long way. Cameras had an immediate positive effect. "As soon as the program started in 2004," Esther told me, "the absentee rate of teachers fell from 40 percent (a very high number) to 20 percent. And magically, now it has been going on since then and that 20 percent is always there." Better yet, students at the camera schools learned more. A year after the program, the kids at camera schools scored substantially higher than their counterparts on standardized tests and were 40 percent more likely to matriculate to regular schools.

Randomized trials are being embraced now in country after country to evaluate the impact of all kinds of public policies. Randomized tests in Kenya have shown the powerful impact of de-worming programs. And randomized tests in Indonesia have shown that a threat of ex post auditing can substantially increase the quality of road construction.

But by far the most important recent randomized social experiment of development policy is the Progresa Program for Education Health and Nutrition. Paul Gertler, one of the six researchers enlisted to evaluate the experiment, told me about how Mexican President Ernesto Zedillo created the program. "Zedillo, who was elected in 1995, was the accidental president," Gertler said. "The original candidate was assassinated and Zedillo, who was the education minister and more of a technocrat, became president. He decided that he wanted to have a major effect on Mexico's poverty and together with members of his administration, he came up with a very unique poverty alleviation program, which is Progresa."

Progresa is a *conditional* transfer of cash to poor people. "To get the cash," Gertler said, "you had to keep your kids in school. To get the cash you had to go to get prenatal care if you are pregnant. You had to go for nutrition monitoring. The idea was to break the intergenerational transfer of poverty because children who typically grow up in poverty tend to remain poor."

Conditioning cash on responsible parenting was a radical idea. And the money would only go to mothers, because Zedillo believed studies suggesting that mothers were more likely than fathers to use the money for their children. Zedillo thought the program had to have a sustained implementation if it had any hope of improving the health and education of children when they became adults. It is not something you would accomplish in just a year.

Zedillo's biggest problem was to try to structure Progresa so that it might outlive his presidency. "Now, the politics in Mexico were such that poverty programs in the past typically changed every presidential campaign," Gertler said. "As hard as this is to believe, the person who was running said that what the current government was doing wasn't very good and it needed to be completely changed. And this is true even if it was the same political party. So, typically what happened is there would be a new president who would come in after five or six years and he would immediately scrap the policies of the previous government and start over again."

Zedillo initially wanted to try Progresa on three to five million families. But he was afraid that he didn't have time. Gertler continued, "If you have a five-year administration and it takes three years to get a program up and running, then it doesn't have much time to have an impact before the new government comes in and closes it." So Zedillo decided instead to conduct a much smaller, but statistically still very large, randomized study of more than 500 villages. On the smaller scale, he could get the program up and running in just a year. He chose to have the program evaluated by independent international academics. It was very much a demonstration project. "Zedillo hoped," Gertler said, "if the evaluation demonstrated that the program had a high benefit-cost ratio, then it would be very hard for the next government to ignore and to close down this program."

So starting in 1997, Mexico began a randomized experiment on more than 24,000 households in 506 villages. In villages assigned to the Progresa program, the mothers of poor families were eligible for three years of cash grants and nutritional supplements if the children made regular visits to health clinics and attended school at least 85 percent of the time. The cash payments were set at roughly two-thirds the wages that children could earn on the free market (and thus increased with the children's age).

The Progresa villages almost immediately showed substantial improvements in education and health. Progresa boys attended school 10 percent more than their non-Progresa counterparts. And school enrollment for Progresa girls was 20 percent higher than for the control group. Overall, teenagers went to school about half a year longer over the initial two-year evaluation period and Progresa students were much less likely to be held back.

The improvements in health were even more dramatic. The program produced a 12 percent lower incidence of serious illness and a 12.7 percent reduction in hemoglobin measures of anemia. Children in the treated villages were nearly a centimeter taller than their non-Progresa peers. A centimeter of additional growth in such a short time is a big deal as a measure of increased health. Gertler explained there were three distinct reasons for the dramatic increase in size: "First, there were nutrition supplements, which were directly given to young kids who are stunted. Second, there was good prenatal and postnatal care to reduce the infection rate. And third, there was just more money to buy food in general."

Sometimes the mechanisms for improvement were more surprising. The evaluators found that birth weights in the Progresa villages increased about 100 grams and that the proportion of low-weight babies dropped by several percentage points. "This is huge in terms of magnitude," Gertler said. The puzzle was why. Pregnant women in the Progresa villages weren't eating better or going for more prenatal visits than pregnant women in the non-Progresa villages. The answer seems to be that Progresa women were more demanding. Gertler explains: "Progresa villages had these sessions where women are told that if you are pregnant when you go for prenatal care here is what to expect. They should weigh and measure you. They should check for anemia. They should check for diabetes. And so it started empowering women and gave them the means and the information to start demanding services that they should get. And then when we interviewed physicians—the physicians say, 'Oh, those Progresa women. They are so much trouble. They come in, they demand more services. They want to talk about everything. We ended up spending so much more time with them. They are really difficult.' "

The Progresa program has proven extremely popular. It has a 97 percent take-up rate. Instead of asking for sacrifice for speculative future benefits, Progresa gives desperately poor mothers money today if they're willing to invest in their children's future. And Zedillo's hope that a demonstration project would tie the hands of his successor worked like a charm. After the 2000 election, when Vicente Fox became president (and for the first time in Mexican history an incumbent president of an opposition party peacefully surrendered power), he was hard put to ignore the Progresa successes in health and education.

"And after we presented our evaluation to them," Gertler said, "the Fox administration came back and said, 'You know, I'm sorry, we need to close down the program. But we are going to open this new program to replace Progresa and this new program is called Oportunidades. It is going to have the same beneficiaries, the same benefits, same management structure, but we have a better name for the program.' " The transparency and third-party verification of the randomized evidence was critical in convincing the government to continue the program. Zedillo's ploy really solved the political economy problem in a big way.

In 2001, Mexico expanded the Progresa (Oportunidades) program to urban areas, adding two million families. Its budget in 2002 was $2.6 billion or about 0.5 percent of Mexican GDP. Progresa is a prime example of how Super Crunching is having real-world impacts. The Super Crunching phenomena often involves large datasets, quick analysis, and the possibility of scaling the results to a much larger population. The Progresa program illustrates the possibility of all three aspects of this new phenomenon. Information on more than 24,000 families was collected, analyzed, and in the space of just five years scaled to cover about 100-fold more people. There has never been a randomized experiment that has had so large a macroeconomic impact.

And the randomization method has revolutionized the way Mexican policy gets made. "The impact of Progresa evaluation has just been huge," Gertler said. "It caused Congress to pass a law which says now all social programs have to be evaluated and it becomes part of the budgetary process. And so now they are evaluating nutrition programs and job programs and microfinance programs and education programs. And it has become the lexicon of the debate over good public policy. The debate is now filled with facts about what works."

What's more, the Progresa idea of conditional cash transfers is spreading like wildfire around the globe. Gertler grows animated as he talks about the impact of the study: "With Progresa we proved that randomized evaluation is possible on a large scale and that the information was very useful for policy and decision making. Progresa is why now thirty countries worldwide have conditional cash transfer programs." Even New York City is now actively thinking about whether it should adopt conditional cash transfers. The Progresa experiment has shown that pragmatic programs can help desperately poor children literally grow.

The Progresa method of randomized trials is also propagating. After Gertler got involved in evaluating the Progresa program, he was asked by the World Bank to be their chief economist for human development. He told me recently, "I spent the last three years helping the World Bank build a capacity of Progresa-like operations, evaluations in 100-plus activities that they were doing. And so we use that as a model to scale up worldwide. It is now entrenched in the World Bank and it is really spreading to other countries."

Mark Twain once said, "Facts are stubborn things, but statistics are more pliable." Government programs like Progresa and software programs like Offermatica show, however, the simple power—the non-pliability, if you will—of randomized trials. You flip coins, put similar people in different treatment groups, and then just look to see what happened to the different groups. There is a purity to this kind of Super Crunching that is hard for even quantiphobes to ignore. Gertler puts it this way: "Randomization just strips away all the roadblocks that you might have or at least it makes them bare. If the people are going to make a political decision, then they are going to make it despite the facts."

In some ways, random trials seem too simple to be part of the Super Crunching revolution. But we have seen that they partake to varying degrees of the same elements. Randomized trials are taking place on larger and larger pools of subjects. CapOne thinks nothing of sending randomized solicitations to hundreds of thousands of prospects. And Offermatica shows how Internet automation has collapsed the period between testing and implementation. Never before has it been possible to test and recalibrate policy in so short a time. But most importantly we've seen how randomization can impact data-driven decision making. The pellucid simplicity of a randomized trial is hard for even political adversaries to resist. Something as seemingly inconsequential as flipping a coin can end up having a massive effect on how the world operates.

CHAPTER 4

How Should Physicians Treat Evidence-Based Medicine?

In 1992, two Canadian physicians from Ontario's McMaster University, Gordon Guyatt and David Sackett, published a manifesto calling for "evidence-based medicine" (EBM). Their core idea was simple. The choice of treatments should be based on the best evidence and, when available, the best evidence should come from statistical research. Guyatt and Sackett didn't call on doctors to be exclusively guided by statistical studies. Indeed, Guyatt is on record as saying that statistical evidence "is never enough." They just wanted statistical evidence to play a bigger role in treatment decisions.

The idea that doctors should give special emphasis to statistical evidence remains controversial to this day. This struggle over EBM parallels the struggle over Super Crunching more generally. Super Crunching is crucially about the impact of statistical analysis on real-world decisions.

The debate over EBM is in large part a debate about whether statistics should impact real-world treatment decisions.

The statistical studies of EBM use the two core Super Crunching techniques. Many of the studies estimate the kind of regression equations that we saw in the first chapter, often with supersized datasets—with tens, and even hundreds, of thousands of subjects. And of course, many of the studies continue to exploit the power of randomization—except now the stakes are much higher. Because of the success of the EBM movement, the pace at which some doctors incorporate results into treatment decisions has accelerated. The Internet's advances in information retrieval have spurred a new technology of influence, and the speed at which new evidence drives decisions has never been greater.

100,000 Lives

Empirical tests of medical treatments have been around for more than a hundred years. As early as the 1840s, the great Austrian physician Ignaz Semmelweis completed a detailed statistical study of maternity clinics in Vienna. As an assistant professor on the maternity ward of the Vienna General Hospital, Semmelweis noticed that women examined by student doctors who had not washed their hands after leaving the autopsy room had very high death rates. When his friend and colleague Jakob Kolletschka died from a scalpel cut, Semmelweis concluded that childbirth (puerperal) fever was contagious. He found that mortality rates dropped from 12 percent to 2 percent if doctors and nurses at the clinics washed their hands in chlorinated lime before seeing each patient.

This startling result, which ultimately would give rise to the germ theory of disease, was fiercely resisted. Semmelweis was ridiculed by other physicians. Some thought his claims lacked a scientific basis because he didn't offer a sufficient explanation for why hand-washing would reduce death. Physicians refused to believe that they were causing their patients' deaths. And they complained that hand-washing several times a day was a waste of their valuable time. Semmelweis was eventually fired. After a nervous breakdown, he ended up in a mental hospital, where he died at the age of forty-seven.

The tragedy of Semmelweis's death, as well as the needless deaths of thousands of women, is ancient history. Doctors today of course know the importance of cleanliness. Medical dramas show them meticulously scrubbing in for operations. But the Semmelweis story remains relevant. Doctors still don't wash their hands enough. Even today, physicians' resistance to hand-washing is a deadly problem. But most importantly, it's still a conflict that is centrally about whether doctors are willing to change their modus operandi because a statistical study says so. This is a conflict that has come to obsess Don Berwick.

A pediatrician and president of the Institute for Healthcare Improvement, Don Berwick inspires some pretty heady comparisons. The management guru Tom Peters calls him "the Mother Teresa of health safety." To me, he is the modern-day Ignaz Semmelweis. For more than a decade, Berwick has been crusading to reduce hospital error. Like Semmelweis, he focuses on the most basic back-end results of our health care system: who lives and who dies. Like Semmelweis, he has tried to use the results of EBM to suggest simple reforms.

Two very different events in 1999 turned Berwick into a crusader for system-wide change. First, the Institute of Medicine published a massive report documenting widespread errors in American medicine. The report estimated that as many as 98,000 people died each year in hospitals as a result of preventable medical errors.

The second event was much more personal. Berwick's own wife, Ann, fell ill with a rare autoimmune disorder of the spinal cord. Within three months she went from completing a twenty-eight-kilometer cross-country ski race in Alaska to barely being able to walk.

The Institute of Medicine report had already convinced Berwick that medical errors were a real problem. But it was his wife's slovenly hospital treatment that really opened Berwick's eyes. New doctors asked the same questions over and over again and even repeated orders for drugs that had already been tried and proven unsuccessful. After her doctors had determined that "time was of the essence" for using chemotherapy to slow deterioration of her condition, Ann had to wait sixty hours before the first dose was finally administered. Three different times, Ann was left on a gurney at night in a hospital subbasement, frightened and alone.

"Nothing I could do...made any difference," Don recalls. "It nearly drove me mad." Before Ann's hospitalization, Berwick was concerned. "Now, I have been radicalized." No longer could he tolerate the glacial movement of hospitals to adopt simple Semmelweis policies to reduce death. He lost his patience and decided to do something about it.

In December 2004, he brazenly announced a plan to save 100,000 lives over the next year and a half. The "100,000 Lives Campaign" challenged hospitals to implement six changes in care to prevent avoidable deaths. He wasn't looking for subtle or sophisticated changes. He wasn't calling for increased precision in surgical operations. No, like Semmelweis before him, he wanted hospitals to change some of their basic procedures. For example, a lot of people after surgery develop lung infections while they're on ventilators. Randomized studies showed that simply elevating the head of the hospital bed and frequently cleaning the patient's mouth substantially reduces the chance of infection. Again and again, Berwick simply looked at how people were actually dying and then tried to find out whether there was large-scale statistical evidence showing interventions that might reduce these particular risks. EBM studies also suggested checks and rechecks to ensure that the proper drugs were prescribed and administered, adoption of the latest heart attack treatments, and use of rapid response teams to rush to a patient's bedside at the first sign of trouble. So these interventions also became part of the 100,000 Lives Campaign.

Berwick's most surprising suggestion, however, is the one with the oldest pedigree. He noticed that thousands of ICU patients die each year from infections after a central line catheter is placed in their chests. About half of all intensive care patients have central line catheters, and ICU infections are deadly (carrying mortality rates of up to 20 percent). He then looked to see if there was any statistical evidence of ways to reduce the chance of infection. He found a 2004 article in *Critical Care Medicine* that showed that systematic hand-washing (combined with a bundle of improved hygienic procedures such as cleaning the patient's skin with an antiseptic called chlorhexidine) could reduce the risk of infection from central-line catheters by more than 90 percent. Berwick estimated that if all hospitals just implemented this one bundle of procedures, they might be able to save as many as 25,000 lives per year. Just as numerical analysis had informed Ignaz so many years before, it was a statistical study that showed Berwick a way to save lives.

Berwick thinks that medical care could learn a lot from aviation, where pilots and flight attendants have a lot less discretion than they used to have. He points to FAA safety warnings that have to be read word for word at the beginning of each flight. "The more I have studied it, the more I believe that less discretion for doctors would improve patient safety," he says. "Doctors will hate me for saying that."

Berwick has crafted a powerful marketing message. He tirelessly travels and is a charismatic speaker. His presentations at times literally sound like a revival meeting. "Every single person in this room," he told one gathering, "is going to save five lives during the forum." He constantly analogizes to real-world examples to get his point across. His audiences have heard him compare health care to the escape of forest-fire jumpers, his younger daughter's soccer team, Toyota, the sinking of a Swedish warship, the Boston Red Sox, Harry Potter, NASA, and the contrasting behaviors of eagles and weasels.

And he is fairly obsessed with numbers. Instead of amorphous goals, his 100,000 Lives Campaign was the first national effort to save a specific number of lives in a set amount of time. The campaign's slogan is "Some Is Not a Number, Soon Is Not a Time."

The campaign signed up more than 3,000 hospitals representing about 75 percent of U.S. hospital beds. Roughly a third of the hospitals agreed to implement all six changes, and more than half used at least three. Before the campaign, the average mortality rate for hospital admits in the United States was about 2.3 percent. For the average hospital in the campaign with 200 beds and about 10,000 admits a year, this meant about 230 annual fatalities. By extrapolating from existing studies, Berwick figured that participating hospitals could save about one life for every eight hospital beds—or about twenty-five lives a year for a 200-bed hospital.

The campaign required participating hospitals to provide eighteen months of mortality data before they began participating and to report updates on a monthly basis of how many people died during the course of the experiment. It's hard to assess at a single hospital with 10,000 admits whether any mortality decrease is just plain luck. Yet when the before-and-after results of 3,000 hospitals are crunched, it's possible to come to a much more accurate assessment of the aggregate impact.

And the news was great. On June 14, 2006, Berwick announced that the campaign had exceeded its goal. In just eighteen months, the six reforms prevented an estimated 122,342 hospital deaths. This precise number can't really be trusted—in part because many hospitals were independently making progress on the problem of preventable medical error. Even without the campaign, it's probable that some of the participating hospitals would have changed the way they do business and saved the lives.

Still, any way you slice it, this is a huge victory for evidence-based medicine. You see, the 100,000 Lives Campaign is centrally about Super Crunching. Berwick's six interventions didn't come from his intuitions, they came from statistical studies. Berwick looked at the numbers to find out what was actually causing people to die and then looked for interventions that had been statistically shown to reduce the risk of those deaths.

But this is statistics on steroids. Berwick succeeded in scaling his campaign to impact two out of every three hospital beds in the country. And the sheer speed of the statistical influence is staggering: saving 100,000 lives in a little more than 500 days. It shows the possibility of quickly going from publication to mass implementation. Indeed, the central-line study was published just two months before the 100,000 Lives Campaign began.

"Don Berwick should win the Nobel Prize for Medicine," says Blair Sadler, head of San Diego Children's Hospital. "He has saved more lives than any doctor alive today." And he's not done either. In December 2006, his Institute for Healthcare Improvement announced the 5 Million Lives Campaign, a two-year initiative to protect patients from 5 million incidents of medical harm. The success of the 100,000 Lives Campaign underscores the potential for translating EBM results into mass action by health care providers.

Old Myths Die Hard

But the continuing problem of dirty hands underscores the difficulty of getting the medical community to follow where the statistics lead them. Even when statistical studies exist, doctors are often blissfully unaware of—or, worse yet, deliberately ignore—statistically prescribed treatments just because that's not the way they were taught to treat. Dozens of studies dating back to 1989 found little support for many of the tests commonly included in a typical annual physical for symptomless people. Routine pelvic, rectal, and testicular exams for those with no symptoms of illness haven't made any difference in overall survival rates. The annual physical exam is largely obsolete. Yet physicians insist on doing them, and in very large numbers.

Dr. Barron Lerner, an internist at Columbia University's College of Physicians and Surgeons, asks patients to come in every year and always listens to their heart and lungs, does a rectal exam, checks lymph nodes, and palpates their abdomens.

"If a patient were to ask me, 'Why are you listening to my heart today?' " he said, "I couldn't say, 'It's going to help me predict whether you will have a heart attack.'

"It's what I was taught and it's what patients have been taught to expect," he said.

Worse yet, there is an extensive literature on "medical myths" that persist in practice long after they've been refuted by strong statistical evidence. On the ground, many practicing doctors still believe:

Vitamin B_{12} deficiencies must be treated with shots because vitamin pills are ineffective.

Patching the eye improves comfort and healing in patients with corneal abrasions.

It is wrong to give opiate analgesics to patients with acute abdomen pain because narcotics can mask the signs and symptoms of peritonitis.

However, there is evidence from carefully controlled randomized trials that each of these beliefs is false. It's not surprising that the untrained general public clings to "folk wisdom" and unproven alternative medicines. But how do medical myths persist among practicing physicians?

Part of the persistence comes from the idea that new studies aren't really needed. There's that old Aristotelian pull. The whole idea of empirical testing goes against the Aristotelian approach that has been a guiding principle of research. Under this approach, researchers should first try to understand the nature of the disease. Once you understand the problem, a solution will become self-evident. Semmelweis had trouble publishing his results because he couldn't give an Aristotelian explanation for why hand-washing saved lives. Instead of focusing on the front-end knowledge about the true nature of disease, EBM shows the power of asking the back-end question of whether specific treatments work. This doesn't mean that medical research should say goodbye to basic research or the Aristotelian approach. But for every Jonas

Salk, we need a Semmelweis and, yes, a Berwick to make sure that the medical profession's hands are clean.

The Aristotelian approach can go seriously wrong if doctors embrace a mistaken conception or model for how an ailment operates. Many doctors still cling to mistaken treatments because they have adopted a mistaken model. As Montana physician and medical myth-buster Robert Flaherty puts it, "It makes pathophysiologic sense, so it must be true."

Moreover, once a consensus has developed about how to treat a particular disease, there is a huge urge in medicine to follow the herd. Yet blindly playing follow the leader can doom us to going the wrong way if the leaders are poorly informed. Here's how it happens: we now know that B_{12} pills are just as effective as shots—they both work about 80 percent of the time. But imagine what happens if, just by chance, the first few patients who are given B_{12} pills don't respond to the treatment positively (while the first few that receive B_{12} shots do). The pioneering clinician will respond to this anecdotal evidence by telling her students that B_{12} pills aren't effective supplements. Subsequent doctors just keep giving the shots, which are in fact effective 80 percent of the time but which are also a lot more expensive and painful. *Voilà,* we have an informational cascade, where an initial mistaken inference based on an extremely small sample is propagated in generation after generation of education.

Other physicians reject evidence-based medicine, saying that there often aren't any good statistical studies to guide their decision. Even today, many medical procedures do not have systematic data to support them. Lisa Sanders, the "Diagnosis" columnist for the *New York Times Sunday Magazine* and an internist at Waterbury Hospital in Connecticut, puts it this way: "Even though evidence-based medicine has made tremendous strides," she says, "still only a tiny fraction of what we do in the course of a day is actually evidence-based."

Lisa is the perfect person to step back and think about the rise of evidence-based medicine, because she has lived through the sea change in emphasis from the early days of her training in the 1990s. "Statistics really wasn't taught until a few years ago," she says. "I was on the beginning of the wave."

Born and raised in South Carolina, Lisa is a true renaissance woman. In a former life before becoming a doctor, she was an Emmy-winning producer for CBS news. Her current "Diagnosis" column is the inspiration for the hit TV show *House,* where she is the lead medical consultant. She cares passionately about evidence-based medicine, but sees its limitations.

"Right now I am teaching all of these residents all the stuff of the physical exam," she tells me with characteristic candor, "but I don't tell them how sensitive it is or how specific it is or what its positive predictive value is because mostly we don't know. I try to—to whatever extent I know about it. But it is still a huge gray area, and when you think that one of your fundamental tools—the physical exam and taking the history—is not at all evidence-based at this point, it's just kind of crazy. So that is where we are. We are just beginning."

In some ways, it shouldn't be surprising that there aren't always evidence-based answers. Medical research is damned expensive. The cost not only cuts down the quantity done, it also biases what kinds of questions get studied and reported. Pharmaceutical companies fund efforts to prove that their drugs are effective, but are not about to ask whether non-proprietary treatments are equally as good.

And even when rigorous empiricism exists, it may not provide information that is relevant to the patient in question. The protocols for clinical trials routinely disqualify patients with many comorbid diseases so that the study can exclusively assess the impact of a treatment on a particular disease. And there has been an unfortunate tradition of excluding women and minorities from these trials.

It's also reasonable for doctors to resist statistical studies that are badly done. Some studies don't ask the right questions or control for sufficient variables. A few years back, a study concluded that excessive caffeine consumption increased the risk of heart disease—without controlling for whether the patients in the study smoked. Turns out that smokers were more likely to drink a lot of coffee and that smoking, not caffeine, was the real cause for the heart failure. This kind of resistance is completely consonant with evidence-based medicine. EBM only asks doctors to assess the quality of different types of evidence and to give high-quality, systematic research the appropriate weight when making their treatment decisions. The problem is that many doctors fail to make any assessment of relevant EBM results because they never learn that a study exists.

"You Could Look It Up"

A physician's decisions can't be driven by statistical results if the physician doesn't know what the statistical result is. For statistical analysis to have impact, there needs to be some kind of a transmission mechanism that disseminates the analysis to the decision makers. The rise of Super Crunching is often accompanied and nurtured by improvements in the technology of transmission so that decision makers can more quickly access and respond to data in real time. We've even seen automation of this transmission link in randomized Internet testing applications. Google's AdWords not only reports your results in real time, but it will automatically shift your web page image toward the version that produces the best results. The faster Super Crunching results are available, the faster they can change a decision maker's choices.

In sharp contrast, the practice of medicine before the EBM movement was shackled with an extremely slow and inefficient mechanism for disseminating medical results. The Institute of Medicine estimates that it took "an average of 17 years for new knowledge generated by randomized controlled trials to be incorporated into practice, and even then application [was] highly uneven." Progress in medical science occurred one funeral at a time. If doctors didn't learn about something in medical school or in their residency, there was a good chance they never would.

As in other Super Crunching contexts, the EBM movement has tried to shorten the time it takes to disseminate important results. The central, but probably most deeply resisted, demand of EBM is the call for physicians to research the problems of their particular patients. Just as Semmelweis angered doctors by calling on them to wash their hands several times a day, the evidence-based medicine movement has had the impertinence to ask doctors to change what they do with their time.

Of course, doctors shouldn't do patient-specific research for all their patients. It would be a huge waste of time to hit the books when a patient presents with the classic symptoms of a common cold. And in the emergency room, there simply isn't time to do research. But academics who have "shadowed" practicing physicians find that about one in three new patients presents questions that would benefit from a review of current research. And this proportion rises for patients newly admitted to a hospital. Yet very few physicians in these studies actually take the time to look up the answers.

The critics of evidence-based medicine often focus on the lack of information. They claim that in many instances high-quality statistical studies just don't exist to provide guidance for the myriad of questions that arise in day-to-day clinical decision making. A deeper reason for resistance is just the opposite problem: there is too much evidence-based information for any individual practitioner to reasonably absorb. On coronary heart disease alone, there are more than 3,600 statistical articles published each year. A specialist who wants to keep up in the field would have to read more than ten articles every day (including weekends). At fifteen minutes an article, that's two and a half hours a day dedicated to reading about just one class of disease. "And it would be a waste of time," Lisa says. "Most of those articles aren't any good."

Clearly, asking physicians to devote such a large amount of time to sifting through mountains of statistical studies was never going to be feasible. From the very beginning, the advocates of evidence-based medicine understood that *information retrieval* technology was critical to allowing practicing physicians to pull the relevant and high-quality information from the massive and ever-changing database of medical research. Indeed, in their original 1992 article, Guyatt and Hackett looked into their crystal ball and imagined how a junior medical resident might respond to "a 43-year-old previously well man who experienced a witnessed grand mal seizure." Instead of just asking a senior resident or attending physician what to do, the evidence-based practitioner would proceed "to the library and, using the 'Grateful Med program,' conduct...a computerized literature search.... The search costs the resident $2.68, and the entire process (including the trip to the library and the time to make a photocopy of the article) [would take] half an hour." The idea that a resident could go to the library, locate, and photocopy an article in half an hour in retrospect was ludicrously optimistic, but something has happened since 1992 to make the idea of half-hour research much less of a pipe-dream: Al Gore's friend, the Internet. It's sometimes hard to remember, but the first web browser wasn't released until 1993.

"It was technology—the computer, the Internet—that finally made it possible for the doctors in the trenches, doctors taking care of patients, to systematically practice evidence-based medicine," Lisa says. "We all now have access to the newest and best in medical research right at our own desks. We can find out quickly, for example, that arthroscopic knee surgery for osteoarthritis and postmenopausal hormone replacement for the prevention of heart disease have lost their standing as effective therapies." Because of the web, every examination room now can have a virtual library.

The retrieval technology of the web has made it a lot easier for physicians to find the results that relate to the specific problems of specific patients. Even though there are more high-quality statistical articles than ever, it is simultaneously faster than ever for physicians to find the haystack's needle. A host of computer-assisted search engines now exist that are dedicated to putting doctors in touch with relevant statistical studies. Infotriever, DynaMed, and FIRSTConsult are all there to help doctors find the summaries of cutting-edge research with just a few clicks of their mouse.

These summaries of results usually have a web link so that the physician can click through and see both the underlying study and all subsequent studies that have cited it. But even without clicking, a doctor can tell a lot from the initial search result just by looking at the "Level of Evidence" designation. Today, every study is given a grade (on a fifteen-category scale developed by the Oxford Center for Evidence-Based Medicine) as a shorthand way to inform the reader of the quality of evidence. The highest possible grade ("1 a") is only awarded when there are multiple randomized trials with similar results, while the lowest grade goes to suggested treatments that are based solely on expert opinion.

This change toward succinctly characterizing the quality of the evidence may well be one of the most important impacts of the evidence-based medicine movement. Now a practitioner in evaluating a study recommendation can have a much better idea of how much he or she can trust the advice. One of the coolest things about Super Crunching regressions is that they not only make predictions but are able to simultaneously tell you how precise the prediction is. The same kind of thing is now happening with these levels of evidence disclosures. EBM not only makes treatment recommendations but simultaneously tells the physician the quality of the data backing up the recommendation.

This grading of evidence is a powerful response to the naysayers who claim that an evidence-based approach won't fly because there simply aren't enough studies to answer all the questions that doctors need to answer. Grading still lets experts answer pressing questions of the day even in the absence of authoritative statistical evidence. It just requires them to reveal the limits of the field's current knowledge. The level of evidence-grading scale is also a simple but real advance in information retrieval. The harried physician can now cruise through dozens of Internet search results and distinguish anecdotes from robust multi-study outcomes.

The openness of the Internet is even transforming the culture of medicine. The results of regressions and randomized trials are out and available not just for doctors but for anyone who has time to Google a few keywords. Doctors are feeling pressured to read not just because their (younger) peers are telling them to, but because increasingly they read to stay ahead of their patients. Just as car buyers are jumping on the Internet before they visit the showroom, many patients are going to sites like Medline to figure out what might be ailing them. The Medline website was originally intended for physicians and researchers. Now, more than a third of its visitors come from the general public.

And Medline has responded by adding twelve consumer health journals and MedlinePlus, a sister site specifically for patients. The Internet is thus not just changing the mechanism by which information is disseminated to physicians, it is changing the technology of influence, the mechanisms by which patients can affect what their physicians do.

Technology can be critical for Super Crunching to change real-world decisions. When a decision maker in business or government commissions a study, the mechanism for transmitting the results is usually not an issue. In that case, the "technology" might be as direct as handing a copy of the results to your boss. But when there are hundreds and even thousands of unsolicited studies on the general topic of interest, the question of how to quickly retrieve the most relevant result often will determine whether the result will even have a chance of altering a decision.

The Future Is Now

The success of evidence-based medicine is the rise of data-based decision making par excellence. It is decision making based not on intuition or personal experience, but on systematic statistical studies. It is Super Crunching that reversed conventional wisdom and found that beta blockers can actually help patients with heart failure. It is Super Crunching that showed that estrogen therapy does not help aging women. And it is Super Crunching that led to the 100,000 Lives Campaign.

So far, the rise of data-based decision making in medicine has largely been limited to questions of treatment. The next wave will almost certainly concern diagnosis.

The database of information that we call the Internet is already having a bizarre impact on diagnosis. The *New England Journal of Medicine* published a description of rounds at a New York teaching hospital. A "fellow in allergy and immunology presented the case of an infant with diarrhea; an unusual rash ('alligator skin'); multiple immunologic abnormalities, including low T-cell function; tissue eosinophilia (of the gastric mucosa) as well as peripheral eosinophilia; and an apparent X-linked genetic pattern (several male relatives died in infancy)." The attending physicians and house staff, after a long discussion, couldn't reach any consensus as to the correct diagnosis. Finally, the professor asked the fellow if she had made a diagnosis, and she reported that she had indeed and mentioned a rare syndrome known as IPEX which fit the symptoms perfectly. When the fellow was asked to explain how she arrived at her diagnosis, she answered: "I entered the salient features into Google, and it popped right up." The attending physician was flabbergasted. "William Osler must be turning over in his grave. You googled the diagnosis? . . . Are we physicians no longer needed?"

The attending's extemporaneous reference to William Osler is particularly apt. Osler, who was one of the founders of Johns Hopkins, is the father of the medical residency program—the continuing cornerstone of all clinical training. Osler would be turning in his grave at the thought of Google diagnoses and Google treatments because the Internet disrupts the dependence of young doctors on the teaching staff as the dominant source of wisdom. Young doctors don't need to defer to the sage experience of their superiors. They can use sources that won't take joy in harassing them.

A bunch of medical schools and private corporations are developing the first generation of "diagnostic-decision support" software. A diagnostic program named "Isabel" allows physicians to enter a patient's symptoms and receive a list of the most likely causes. It will even tell the doctor whether a patient's symptoms might be caused by the use of over 4,000 drugs. The Isabel database associates more than 11,000 specific illnesses with a host of clinical findings, lab results, patient histories, and the symptoms themselves. The Isabel programmers created a taxonomy for all of the diseases and then tutored a database by statistically searching for word patterns in journal articles that were most likely to be associated with each disease. This statistical search algorithm dramatically increased the efficiency of coding particular disease/symptom associations. And it also allows the database to continually update as new articles emerge having a high prediction of relevance. Instead of all-or-nothing Boolean searches, Super Crunching predictions about relevance are crucial to Isabel's success.

The Isabel program grew out of a stockbroker's own experience with suffering caused by misdiagnosis. In 1999, Jason Maude's three-year-old daughter Isabel was misdiagnosed by a London resident as having chicken pox and sent home. It was only the next day when her organs began shutting down that Joseph Britto, an attending intensive care doctor at the hospital, realized that she in fact had a potentially fatal flesh-eating virus. Though Isabel ultimately recovered, her father was so shaken by the experience that he quit his finance job. Maude and Britto together founded a company and started to develop the Isabel software to fight misdiagnosis.

Misdiagnosis accounts for about one-third of all medical error. Autopsy studies show that doctors seriously misdiagnose fatal illnesses about 20 percent of the time. "If you look at settled malpractice claims," Britto said, "diagnosis error is about twice or three times as common as prescription error." The bottom line is that millions of patients are being treated for the wrong disease. And even more troubling, a 2005 editorial in the *Journal of the American Medical Association* concludes that there hasn't been a perceptible improvement in the misdiagnosis rate in the last several decades.

The ambition of Isabel is to change the stagnation in the science of diagnosis. Maude puts it simply: "Computers are better at remembering things than we are." There are more than 11,000 diseases in the world and the human brain is not adept at remembering all the symptoms that give rise to each. Isabel actually markets itself as the Google of medical diagnosis. Like Google, it aids us in searching for and retrieving information from a large database.

The biggest reason for misdiagnosis is "premature closure." Doctors think they have a bead on the correct diagnosis—like the resident's idea that Isabel Maude had chicken pox—and they close their minds to other possibilities. Isabel is a reminder system of other possibilities. It actually produces a page that asks, "Have you considered . . . ?" Just proactively reminding doctors of other possibilities can have profound effects.

In 2003, a four-year-old boy from rural Georgia was admitted to a children's hospital in Atlanta. The boy had been sick for months, with a fever that just would not go away. The doctors on duty that day ordered blood tests, which showed that the boy had leukemia. They ordered a strong course of chemotherapy to start the very next day.

John Bergsagel, a senior oncologist at the hospital, was troubled by light brown spots on the boy's skin that didn't quite fit the normal symptoms of leukemia. Still, Bergsagel had lots of paperwork to get through and was tempted to rely on the blood test that clearly indicated leukemia. "Once you start down one of these clinical pathways," Bergsagel said, "it's very hard to step off."

By chance, Bergsagel had recently seen an article about Isabel and had signed up to be one of the beta testers of the software. So instead of turning to the next case, Bergsagel sat down at a computer and entered the boy's symptoms. Near the top of the "Have you considered . . . ?" list was a rare form of leukemia that chemotherapy does not cure. Bergsagel had never heard of it before, but sure enough, it often presented with brown skin spots.

Researchers have found that about 10 percent of the time, Isabel helps doctors include a major diagnosis that they would not have considered but should have. Isabel is constantly putting itself to the test. Every week the *New England Journal of Medicine* includes a diagnostic puzzler in its pages. Simply cutting and pasting the patient's case history into the input section allows Isabel to produce a list of ten to thirty diagnoses. Seventy-five percent of the time these lists include what the *Journal* (usually via autopsies) verifies as the correct diagnosis. And manually entering findings into more tailored input fields raises Isabel's success rate to 96 percent. The program doesn't pick out a single diagnosis. "Isabel is not an oracle," Britto says. It doesn't even give the probable likelihood or rank the most likely diagnosis. Still, narrowing the likely causes from 11,000 to 30 unranked diseases is substantial progress.

I love the TV show *House*. But the central character, who has unsurpassed powers as a diagnostician, never does any research. He relies on his experience and Sherlockian deductive powers to pull a diagnostic rabbit out of the hat each week. *House* makes excellent drama, but it's no way to run a health care system. I've suggested to my friend Lisa Sanders, who recommends script ideas for the series, that *House* should have an episode in which the protagonist vies against data-based diagnostics—á la Kasparov vs. the IBM computer. Isabel's Dr. Joseph Britto doesn't think it would work. "Each episode would be five or seven minutes instead of an hour," he explains. "I could see Isabel working much better with *Grey's Anatomy* or *ER* where they have to make a lot of decisions under a lot of time pressure." Only in fiction does man beat the machine.

And Super Crunching is going to make the diagnostic predictions even better. At the moment these softwares are still basically crunching journal articles. Isabel's database has tens of thousands of associations but at the end of the day it is solely a compilation of information published in medical journal articles. A team of doctors aided with Google-like natural language searches for published symptoms that have been associated with a particular disease and enters the results into the diagnostic database.

As it currently stands, if you go to see a doctor or are admitted to the hospital, the results of your experience have absolutely no value to our collective knowledge of medicine—save for the exceptional case in which your doctor decides to write it up for a journal or when your case happens to be included in a specialized study. From an information perspective, most of us die in vain. Nothing about our life or death helps the next generation.

The rapid digitalization of medical records means that for the first time ever, doctors are going to be able to exploit the rich information that is embedded in our aggregate health care experience. Instead of giving an undifferentiated list of possible diagnoses, Isabel will, within one or two years, be able to give the likelihood of particular diseases that are conditioned on your particular symptoms, patient history, and test results. Britto grows animated as he describes the possibilities. "You have someone who comes in with chest pains, sweating, palpitations, and is over fifty years old," he says. "You as a doctor might be interested to know that in the last year, at Kaiser Permanente Mid-Atlantic, these symptoms have turned out to be much more often myocardial infarction and perhaps less commonly a dissecting aneurysm."

With digital medical records, doctors don't need to type in symptoms and query their computer. Isabel can automatically extract the information from the records and generate its prediction. In fact, Isabel has recently teamed with NextGen to create a software with a structured flexible set of input fields to capture essential data. Instead of the traditional record keeping where doctors would non-systematically dictate what in retrospect seemed relevant, NextGen collects much more systematic data from the get-go. "I don't like saying this loudly to my colleagues," Britto confides, "but in a sense you engineer this physician out of the role of having to enter these data. If you have structured fields, you then force a physician to go through them and therefore the data that you get are much richer than had you left him on his own to write case notes, where we tend to be very brief."

Super Crunching these massive new databases will give doctors for the first time the chance to engage in real-time epidemiology. "Imagine," Britto said, "Isabel might tell you that an hour ago on the fourth floor of your hospital a patient was admitted who had similar features of infection and blisters." Some patterns are much easier to see in aggregate than from casual observation by individual participants.

Instead of relying solely on expert-filtered data, diagnosis should also be based on the experience of the millions of people who use the health care system. Indeed, database analysis might ultimately lead to better decision making about how to investigate a diagnosis. For people with your symptoms, what tests produced useful information? What questions were most helpful? We might even learn the best order in which to ask questions.

When Britto started learning how to fly an airplane back in 1999, he was struck by how much easier it was for pilots to accept flight support software. "I asked my flight instructor what he thought accounted for the difference," Britto said. "He told me, 'It is very simple, Joseph. Unlike pilots, doctors don't go down with their planes.'"

This is a great line. However, I think physician resistance to evidence-based medicine has much more to do with the fact that no one likes to change the basic way that they have been operating. Ignaz Semmelweis found that out a long time ago when he had the gall to suggest that doctors should wash their hands repeatedly throughout the day. The same reaction is at play when the EBM crowd suggests that doctors should do patient-specific research about the most appropriate treatment. Many physicians have effectively ceded a large chunk of control of treatment choice to Super Crunchers. Lisa Sanders distinguishes diagnosis, which she calls "my end," from the question of the appropriate therapy, which she says "is really in the hands of the experts." When she says "the experts," she means Super Crunchers, the Ph.D.s who are cranking out the statistical studies showing which treatment works best. Very soon, however, Isabel will start to invade the physicians' end of the process. We will see the fight move to evidence-based diagnosis. Isabel Healthcare is careful to emphasize that it only provides diagnostic support. But the writing is on the wall. Structured electronic input software may soon force physicians to literally answer the computer's questions.

The Super Crunching revolution is the rise of data-driven decision making. It's about letting your choices be guided by the statistical predictions of regressions and randomized trials. That's really what the EBM crowd wants. Most physicians (like just about every other decision maker we have and will encounter) still cling to the idea that diagnosis is an art where their expertise and intuition are paramount. But to a Super Cruncher, diagnosis is merely another species of prediction.

CHAPTER 5

Experts Versus Equations

The previous chapters have been awash with Super Crunching predictions. Marketing crunchers predict what products you will want to buy; randomized studies predict how you'll respond to a prescription drug (or a website or a government policy); eHarmony predicts who you'll want to marry.

So who's more accurate, Super Crunchers or traditional experts? It turns out this is a question that researchers have been asking for decades. The intuitivists and clinicians almost ubiquitously argue that the variables underlying their own decision making can't be quantified and reduced to a non-discretionary algorithm. Yet even if they're right, it is possible to test independently whether decision rules based on statistical prediction outperform the decisions of traditional experts who base their decisions on experience and intuition. In other words, Super Crunching can be used to adjudicate whether experts can in fact outpredict the equations generated by regressions or randomized experiments. We can step back and use Super Crunching to test its own strength.

This is just the thought that occurred to Ted Ruger, a law professor at the University of Pennsylvania, as he was sitting in a seminar back in 2001 listening to a technical Super Crunching article by two political scientists, Andrew Martin and Kevin Quinn. Martin and Quinn were presenting a paper claiming that, by using just a few variables concerning the politics of the case, they could predict how Supreme Court justices would vote.

Ted wasn't buying it. Ted doesn't look anything like your usual anemic academic. He has a strapping athletic build with a square chin and rugged good looks (think of a young Robert Redford with dark brown hair). As he sat in that seminar room, he didn't like the way these political scientists were describing their results. "They actually used the nomenclature of prediction," he told me. "I am sitting in the audience as somewhat of a skeptic." He didn't like the fact that all the paper had done was try to predict the past. "Like a lot of legal or political science research," he said, "it was retrospective in nature."

So after the seminar he went up to them with a suggestion. "In some sense, the genesis of this project was my talking to them afterwards and saying, well why don't we run the test forward?" And as they talked, they decided to run a horse race, to create "a friendly interdisciplinary competition" to compare the accuracy of two different ways to predict the outcome of Supreme Court cases. In one corner stood the Super Crunching predictions of the political scientists and in the other stood the opinions of eighty-three legal experts. Their assignment was to predict in advance the votes of the individual justices for every case that was argued in the Supreme Court's 2002 term. The experts were true legal luminaries, a mixture of law professors, practitioners, and pundits (collectively thirty-eight had clerked for a Supreme Court justice, thirty-three held chaired professorships, and five were current or former law school deans). While the Super Crunching algorithm made predictions for all the justices' votes in all the cases, the experts were called upon just to predict the votes for cases in their area of expertise.

Ted didn't think it was really a fair fight. The political scientists' model took into account only six factors: (1) the circuit court of origin; (2) the issue area of the case; (3) the type of petitioner (e.g., the United States, an employer, etc.); (4) the type of respondent; (5) the ideological direction (liberal or conservative) of the lower court ruling; and (6) whether the petitioner argued that a law or practice is unconstitutional. "My initial sense," he said, "was that their model was too reductionist to capture the nuances of the decision making and thus legal experts could do better." After all, detailed knowledge of the law and past precedent should count for something.

This simple test implicates some of the most basic questions of what law is. Justice Oliver Wendell Holmes created the idea of legal positivism by announcing, "The life of the law has not been logic; it has been experience." For Holmes, the law was nothing more than "a prediction of what judges in fact will do." Holmes rejected the view of Harvard's dean (and the champion of the Socratic method for legal education) Christopher Columbus Langdell that "law is a science, and that all the available materials of that science are contained in printed books." Holmes felt that accurate prediction had a "good deal more to do" with "the felt necessities of the time, the prevalent moral and political theories, intuitions of public policy, avowed or unconscious, even the prejudices which judges share with their fellow-men."

The dominant statistical model of political science is Holmesian in that it places almost exclusive emphasis on the judge's prejudices, his or her personal ideological views. Political scientists often assumed these political ideologies to be fixed and neatly arrayed along a single numeric spectrum from liberal to conservative. The decision trees produced by this kind of Super Crunching algorithm are anything but nuanced. Using historical data on 628 cases previously decided by these nine justices, Martin and Quinn first looked to see when the six factors predicted that the decision would be a unanimous affirmance or reversal. Then, they used the same historic cases to find the flowchart (a conditional combination of factors) that best predicted the votes of the individual justices in non-unanimous cases. For example, consider the following flowchart that was used to forecast Justice Sandra Day O'Connor's votes in the actual study:

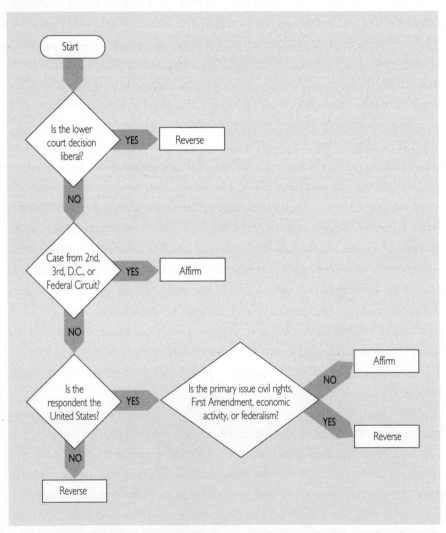

SOURCE: Andrew D. Martin et al., "Competing Approaches to Predicting Supreme Court Decision Making," 2 *Perspectives on Politics* 763 (2004).

This predictive flowchart is incredibly crude. The first decision point predicts that O'Connor would vote to reverse whenever the lower court decision was coded as being "liberal." Hence, in *Grutter v. Bollinger,* the 2002-term case challenging the constitutionality of Michigan Law School's affirmative action policy, the model erroneously forecasted that O'Connor would vote to reverse simply because the lower court's decision was liberal (in upholding the law school's affirmative action policy). With regard to "conservative" lower court decisions, the flowchart is slightly more complicated, conditioning the prediction on the circuit court origin, the type of respondent, and the subject area of the case. Still, this statistical prediction completely ignores the specific issues in the case and the past precedent of the Court. Surely legal experts with a wealth of knowledge about the specific issue could do better.

Notice in the statistical model that humans are still necessary to code the case. A kind of expertise is essential to say whether the lower court decision was "liberal" or "conservative." The study shows how statistical prediction can be made compatible with and dependent upon subjective judgment. There is nothing that stops statistical decision rules from depending on subjective opinions of experts or clinicians. A rule can ask whether a nurse believes a patient looks "hinky." Still, this is a very different kind of expertise. Instead of calling on the expert to make an ultimate prediction, the expert is asked to opine on the existence or absence of a particular feature. The human expert might have some say in the matter, but the Super Crunching equation limits and channels this discretion.

Ted's simple idea of "running the test forward" set the stage for a dramatic test that many insiders watched with interest as it played out during the course of the Court's term. Both the computer and the experts' predictions were posted publicly on a website before the decision was announced, so people could see the results come as opinion after opinion was handed down.

The experts lost. For every argued case during the 2002 term, the model predicted 75 percent of the Court's affirm/reverse results correctly, while the legal experts collectively got only 59.1 percent right. Super Crunching was particularly effective at predicting the crucial swing votes of Justices O'Connor and Kennedy. The model predicted Justice O'Connor's vote correctly 70 percent of the time while the experts' success rate was only 61 percent.

How can it be that an incredibly stripped-down statistical model outpredicted not just lawyers, but experts in the field who had access to detailed information about the cases? Is this result just some statistical anomaly? Does it have something to do with idiosyncrasies or the arrogance of the legal profession? These are the central questions of this chapter. The short answer is that Ted's test is representative of a much wider phenomenon. For decades, social scientists have been comparing the predictive accuracies of Super Crunchers and traditional experts. In study after study, there is a strong tendency for the Super Crunchers to come out on top.

Meehl's "Disturbing Little Book"

Way back in 1954, Paul Meehl wrote a book called *Clinical Versus Statistical Prediction.* This slim volume created a storm of controversy among psychologists because it reported the results of about twenty other empirical studies that compared how well "clinical" experts could predict relative to simple statistical models. The studies concerned a diverse set of predictions, such as how patients with schizophrenia would respond to electroshock therapy or how prisoners would respond to parole. Meehl's startling finding was that none of the studies suggested that experts could outpredict statistical equations.

Paul Meehl was the perfect character to start this debate. He was a towering figure in psychology who eventually became president of the American Psychological Association. He's famous for helping to develop the MMPI (the Minnesota Multiphasic Personality Inventory), which to this day is one of the most frequently used personality tests in mental health. What really qualified Meehl to lead the man-versus-machine debate was that he cared passionately about both sides. Meehl was an experimental psychologist who thought there was value to clinical thinking. He was driven to write his book by the personal conflict between his subjective certainty that clinical experience conferred expertise, and "the disappointing findings on the reliability and validity of diagnostic judgments and prognostications by purported experts."

Because of his book's findings, some people inferred that he was an inveterate number cruncher. In his autobiography, Meehl tells about a party after a seminar where a group of experimental psychologists privately toasted him for giving "the clinicians a good beating." Yet they were shocked to learn that he valued psychoanalysis and even had a painting of Freud in his office. Meehl believed the studies that showed that statisticians could make better predictions about many issues, but he also pointed to the interpretation of dreams in psychoanalysis as a "striking example of an inferential process difficult to actuarialize and objectify." Meehl writes:

> I had not then completed a full-scale analysis but I [told them I] had some 85 couch hours with a Vienna-trained analyst, and my own therapeutic mode was strongly psychodynamic. . . . The glowing warmth of the gathering cooled noticeably. A well-known experimental psychologist became suddenly hostile. He glared at me and said, "Now, come on, Meehl, how could anybody like you, with your scientific training at Minnesota, running rats and knowing math, and giving a bang-up talk like you just gave, how could *you* think there is anything to that Freudian dream shit?"

Meehl continued to inhabit this schizophrenic space for fifty more years. His initial study—which he playfully referred to as "my disturbing little book"—was only an opening salvo in what would become not only a personal lifelong passion, but also a veritable industry of man-versus-machine studies by others.

Researchers have now completed dozens upon dozens of studies comparing the success rates of statistical and expert approaches to decision making. The studies have analyzed the relative ability of Super Crunchers and experts in predicting everything from marital satisfaction and academic success to business failures and lie detection. As long as you have a large enough dataset, almost any decision can be crunched. Studies have suggested that number crunchers using statistical databases can even outperform humans in guessing someone's sexual orientation or in constructing a satisfying crossword puzzle.

Just recently Chris Snijders, a professor at Eindhoven University of Technology, in the Netherlands, decided to see whether he could out-purchase professional corporate buyers. Snijders (rhymes with White Castle "sliders") had collected a database on more than 5,200 computer equipment and software purchases by more than 700 Dutch businesses. For each purchase, Snijders had information on more than 300 aspects of the transaction—including several aspects of purchasing satisfaction such as whether the delivery was late or non-conforming, or whether the product had insufficient documentation. "I had the feeling," he told me, "there is knowledge in there that could be useful somehow to the world. So I went out with my papers trying to explain what is in there to firms, and then they laughed at me. 'What do *you* know? You never worked in any kind of company. Forget your data.' So this was the reason I did the test. If you think that these data say nothing, well, we will see."

Snijders used part of his data to estimate a regression equation relating the satisfaction of corporate purchasers to fourteen aspects of the transaction—things like the size and reputation of the supplier as well as whether lawyers were involved in negotiating the contract. He then used a different set of transactions to run a test pitting his Super Crunched predictions against the predictions of professional purchasing managers. Just as in the Supreme Court test, each of the purchasing experts was presented with a handful of different purchasing cases to analyze.

Just as in the earlier studies, Snijders's purchasing managers couldn't outperform a simple statistical formula to predict timeliness of delivery, adherence to the budget, or purchasing satisfaction. He and his coauthors found that the judgments of professional managers were "meager at best." The Super Crunching formula outperformed even above-average managers. More experienced managers did no better than novices. And managers reviewing transactions in their own industry did not fare any better than those reviewing transactions in different industries. In the end, the study suggests purchasing professionals cannot outpredict even a simple regression equation. Snijders is confident that this is true generally. "As long as you have some history and some quantifiable data from past experiences," he claims, the regression will win. "It's not just pie in the sky," he said. "I have the data to support this."

Snijders's results are just a recent example in a long line of man-versus-machine comparisons. Near the end of his life, Meehl, together with Minnesota protégé William Grove, completed a "meta" analysis of 136 of these man-versus-machine studies. In only 8 of 136 studies was expert prediction found to be appreciably more accurate than statistical prediction. The rest of the studies were equally divided between those where statistical prediction "decisively outperformed" expert prediction, and those where the accuracy was not appreciably different. Overall, when asked to make binary predictions, the average expert in these wildly diverse fields got it right about two-thirds of the time (66.5 percent). The Super Crunchers, however, had a success rate that was almost three-quarters (73.2 percent). The eight studies favoring the expert weren't concentrated in a particular issue area and didn't have any obvious characteristics in common. Meehl and Grove concluded, "The most plausible explanation of these deviant studies is that they arose by a combination of random sampling errors (8 deviant out of 136) and the clinicians' informational advantage in being provided with more data than the actuarial formula."

Why Are Humans Bad at Making Predictions?

These results are the stuff of science fiction nightmares. Our best and brightest experts in field after field are losing out to Super Crunching predictions. How can this be happening?

The human mind tends to suffer from a number of well-documented cognitive failings and biases that distort our ability to predict accurately. We tend to give too much weight to unusual events that seem salient. For example, people systematically overestimate the probability of "newsworthy" deaths (such as murder) and underestimate the probability of more common causes of death. Most people think that having a gun in your house puts your children at risk. However, Steve Levitt, looking at statistics, pointed out that "on average, if you own a gun and have a swimming pool in the yard, the swimming pool is almost 100 times more likely to kill a child than the gun is."

Once we form a mistaken belief about something, we tend to cling to it. As new evidence arrives, we're likely to discount disconfirming evidence and focus instead on evidence that supports our preexisting beliefs.

In fact, it's possible to test your own ability to make unbiased estimates. For each of the following ten questions, give the range of answers that you are 90 percent confident contains the correct answer. For example, for the first question, you are supposed to fill in the blanks: "I am 90 percent confident that Martin Luther King's age at the time of his death was somewhere between __ years and __ years." Don't worry about not knowing the exact answer—and no using Google. The goal here is to see whether you can construct confidence intervals that include the correct answer 90 percent of the time. Here are the ten questions:

	Low	High
1. What was Martin Luther King, Jr.'s age at death?	____	____
2. What is the length of the Nile River, in miles?	____	____
3. How many countries belong to OPEC?	____	____
4. How many books are there in the Old Testament?	____	____
5. What is the diameter of the moon, in miles?	____	____
6. What is the weight of an empty Boeing 747, in pounds?	____	____
7. In what year was Mozart born?	____	____
8. What is the gestation period of an Asian elephant, in days?	____	____
9. What is the air distance from London to Tokyo, in miles?	____	____
10. What is the deepest known point in the ocean, in feet?	____	____

Answering "I have no idea" is not allowed. It's also a lie. Of course you have some idea. You know that the deepest point in the ocean is more than 2 inches and less than 100,000 miles. I've included the correct answers below so you can actually check to see how many you got right. You can't win if you don't play.*

If all ten of your intervals include the correct answer, you're underconfident. Any of us could have made sure that this occurred—just by making our answers arbitrarily wide. I'm 100 percent sure Mozart was born sometime between 33 BC and say, 1980. But almost everyone who answers these questions has the opposite problem of overconfidence—they can't help themselves from reporting ranges that are too

* **The correct answers:** (1) 39 years. (2) 4,187 miles. (3) 13 countries. (4) 39 books. (5) 2,160 miles. (6) 390,000 pounds. (7) 1756. (8) 645 days. (9) 5,959 miles. (10) 36,198 feet.

small. People think they know more than they actually know. In fact, when Ed Russo and Paul Schoemaker tested more than 1,000 people, they found that most people missed between four and seven of the questions. Less than 1 percent of the people gave ranges that included the right answer nine or ten times. Ninety-nine percent of people were overconfident.

So humans not only are prone to make biased predictions, we're also damnably overconfident about our predictions and slow to change them in the face of new evidence.

In fact, these problems of bias and overconfidence become more severe the more complicated the prediction. Humans aren't so bad at predicting simple things—like whether a Coke can that has been shaken will spurt. Yet when the number of factors to be considered grows and it is not clear how much weight will be placed on the individual factors, we're headed for trouble. Coke-can prediction is pretty much one factor: when a can is recently shaken, we can predict almost with certainty the result. In noisy environments, it's often not clear what factors should be taken into account. This is where we tend mistakenly to bow down to experts who have years of experience—the baseball scouts and physicians—who are confident that they know more than the average Joe.

The problem of overconfidence isn't just a problem of academic experiments either. It distorts real-world decisions. Vice President Cheney as late as June 2005 predicted on *Larry King Live* that the major U.S. involvement in Iraq would end before the Bush administration left office. "The level of activity that we see today from a military standpoint, I think will clearly decline," he said with confidence. "I think they're in the last throes, if you will, of the insurgency." The administration's overconfidence with regard to the cost of the war was, if anything, even more extreme. In 2002, Glenn Hubbard, chairman of the president's Council of Economic Advisers, predicted that the "costs of any such intervention would be very small." In April 2003, Defense Secretary Donald Rumsfeld dismissed the idea that reconstruction would be costly: "I don't know," he said, "that there is much reconstruction to do." Several of the key planners were convinced that Iraq's own oil revenue could pay for the war. "There's a lot of money to pay for this that doesn't have to be U.S. taxpayer money," Deputy Defense Secretary Paul Wolfowitz predicted. "And it starts with the assets of the Iraqi people.... We're dealing with a country that can really finance its own reconstruction." While it's easy to dismiss these forecasts as self-interested spin, I think it's more likely that these were genuine beliefs of decision makers who, like the rest of us, have trouble updating our beliefs in the face of disconfirming information.

In contrast to these human failings, think about how well Super Crunching predictions are structured. First and foremost, Super Crunchers are better at making predictions because they do a better job at figuring out what weights should be put on individual factors in making a prediction. Indeed, regression equations are so much better than humans at figuring out appropriate weights that even very crude regressions with just a few variables have been found to outpredict humans. Cognitive psychologists Richard Nisbett and Lee Ross put it this way, "Human judges are not merely worse than optimal regression equations; they are worse than almost any regression equation."

Unlike self-involved experts, statistical regressions don't have egos or feelings. Being "unemotional is very important in the financial world," said Greg Forsythe, a senior vice president at Schwab Equity Ratings, which uses computer models to evaluate stocks. "When money is involved, people get emotional." Super Crunching models, in contrast, have no emotional commitments to their prior predictions. As new data are collected, the statistical formulae are recalculated and new weights are assigned to the individual elements.

Statistical predictions are also not overconfident. Remember Farecast.com doesn't just predict whether an airline price is likely to increase; it also tells you what proportion of the time the prediction is going to be true. Same goes with randomized trials—they not only give you evidence of causation, but they also tell you how good the causal evidence is. Offermatica tells you that a new web design causes 12 percent more sales and that it's 95 percent sure that the true causal effect is between 10.5 percent and 13.5 percent. Each statistical prediction comes with its own confidence interval.

In sharp contrast to traditional experts, statistical procedures not only predict, they also tell you the quality of the prediction. Experts either don't tell you the quality of their predictions or are systematically overconfident in describing their accuracy. Indeed, this difference is at the heart of the shift to evidence-based medical guidelines. The traditional expert-generated guidelines just gave undifferentiated pronouncements of what physicians should and should not do. Evidence-based guidelines, for the first time, explicitly tell physicians the quality of the evidence underlying each suggested practice. Signaling the quality of the evidence lets physicians (and patients) know when a guideline is written in stone and when it's just their best guess given limited information.

Of course, data analysis can itself be riddled with errors. Later on, I'll highlight examples of data-based decision making that have failed spectacularly. Still, the trend is clear. Decisions that are backed by quantitative prediction are at least as good as and often substantially better than decisions based on mere lived experience. The mounting evidence of statistical superiority has led many to suggest that we should strip experts of at least some of their decision-making authority. As Dr. Don Berwick said, physicians might perform better if, like flight attendants, they were forced to follow more scripted procedures.

Why Not Both?

Instead of simply throwing away the know-how of traditional experts, wouldn't it be better to combine Super Crunching and experiential knowledge? Can't the two types of knowledge peacefully coexist?

There is some evidence to support the possibility of peaceful coexistence. Traditional experts make better decisions when they are provided with the results of statistical prediction. Those who cling to the authority of traditional experts tend to embrace the idea of combining the two forms of knowledge by giving the experts "statistical support." The purveyors of diagnostic software are careful to underscore that its purpose is only to provide support and suggestions. They want the ultimate decision and discretion to lie with the doctor. Humans usually do make better predictions when they are provided with the results of statistical prediction. The problem, according to Chris Snijders, is that, even with Super Crunching assistance, humans don't predict as well as the Super Crunching prediction by itself. "What you usually see is the judgment of the aided experts is somewhere in between the model and the unaided expert," he said. "So the experts get better if you give them the model. But still the model by itself performs better." Humans too often wave off the machine predictions and cling to their misguided personal convictions.

When I asked Ted Ruger whether he thought he could outpredict the computer algorithm if he had access to the computer's prediction, he caught himself sliding once again into the trap of overconfidence. "I *should* be able to beat it," he started, but then corrected himself. "But maybe not. I wouldn't really know what my thought process would be. I'd look at the model and then I would sort of say, well, what could I do better here? I would probably muck it up in a lot of cases."

Evidence is mounting in favor of a different and much more demeaning, dehumanizing mechanism for combining expert and Super Crunching expertise. In several studies, the most accurate way to exploit traditional expertise is to merely add the expert evaluation as an additional factor in the statistical algorithm. Ted's Supreme Court study, for example, suggests that a computer that had access to human predictions would rely on the experts to determine the votes of the more liberal justices (Breyer, Ginsburg, Souter, and Stevens)—because the unaided experts outperformed the Super Crunching algorithm in predicting the votes of these justices.

Instead of having the statistics as a servant to expert choice, the expert becomes a servant of the statistical machine. Mark E. Nissen, a professor at the Naval Postgraduate School in Monterey, California,

who has tested computer-versus-human procurement, sees a fundamental shift toward systems where the traditional expert is stripped of
his or her power to make the final decision. "The newest space, and the
one that's most exciting, is where machines are actually in charge," he
said, "but they have enough awareness to seek out people to help them
when they get stuck." It's best to have the man and machine in dia-
logue with each other, but, when the two disagree, it's usually better
to give the ultimate decision to the statistical prediction.

The decline of expert discretion is particularly pronounced in the
case of parole. In the last twenty-five years, eighteen states have replaced their parole systems with sentencing guidelines. And those
states that retain parole have shifted their systems to rely increasingly
on Super Crunching risk assessments of recidivism. Just as your credit
score powerfully predicts the likelihood that you will repay a loan, parole boards now have externally validated predictions framed as numerical scores in formula like the VRAG (Violence Risk Appraisal Guide),
which estimates the probability that a released inmate will commit a
violent crime. Still, even reduced discretion can give rise to serious risk
when humans deviate from the statistically prescribed course of action.

Consider the worrisome case of Paul Herman Clouston. For over
fifty years, Clouston has been in and out of prison in several states for
everything from auto theft and burglary to escape. In 1972, he was
convicted of murdering a police officer in California. In 1994, he was
convicted in Virginia of aggravated sexual battery, abduction, and
sodomy, and of assaulting juveniles in James City County, Virginia.
He had been serving time in a Virginia penitentiary until April 15,
2005, when he was released on mandatory parole six months before the
end of his nominal sentence.

As soon as Clouston hit the streets, he fled. He failed to report for
parole and failed to register as a violent sex offender. He is now one of
the most wanted men in Virginia. He is on the U.S. Marshals' Most
Wanted list and was recently featured on *America's Most Wanted*. But
why did this seventy-one-year-old, who had served his time, flee and
why did he make all of these most wanted lists?

The answer to both questions is the SVPA. In April of 2003, Virginia became the sixteenth state in our nation to enact a "Sexually Violent Predator Act" (SVPA). Under this extraordinary statute, an offender, after serving his full sentence, can be found to be a "sexually violent predator" and subject to civil commitment in a state mental hospital until a judge is satisfied he no longer presents an undue risk to public safety.

Clouston probably fled because he was worried that he would be adjudged to be a sexual predator (defined in the statute as someone "who suffers from a mental abnormality or personality disorder which makes the person likely to engage in the predatory acts of sexual violence"). And the state made Clouston "most wanted" for the very same reason.

The state was also embarrassed that Clouston had ever been released in the first place. You see, Virginia's version of the SVPA contained a Super Crunching innovation. The statute itself included a "tripwire" that automatically sets the commitment process in motion if a Super Crunching algorithm predicted that the inmate had a high risk of sexual offense recidivism. Under the statute, commissioners of the Virginia Department of Corrections were directed to review for possible commitment all prisoners about to be released who, and I'm quoting the statute here, "receive a score of four or more on the Rapid Risk Assessment for Sexual Offender Recidivism." The Rapid Risk Assessment for Sexual Offender Recidivism (RRASOR) is a point system based on a regression analysis of male offenders in Canada. A score of four or more on the RRASOR translates into a prediction that the inmate, if released, would in the next ten years have a 55 percent chance of committing another sex offense.

The Supreme Court in a 5–4 decision has upheld the constitutionality of prior SVPAs—finding that indefinite *civil* commitment of former inmates does not violate the Constitution. What's amazing about the Virginia statute is that it uses Super Crunching to trigger the commitment process. John Monahan, a leading expert in the use of risk-assessment instruments, notes, "Virginia's sexually violent predator statute is the first law ever to specify, in black letter, the use of a named actuarial prediction instrument *and an exact cut-off score* on that instrument."

Clouston probably never should have been released because he had a RRASOR score of four. The state has refused to comment on whether they failed to assess Clouston's RRASOR score as directed by the statute or whether the committee reviewing his case chose to release him notwithstanding the statistical prediction of recidivism. Either way, the Clouston story seems to be one where human discretion led to the error of his release.

It was a mistake, that is, if we trust the RRASOR prediction. Before rushing to this conclusion, however, it's worthwhile to look at what exactly qualified Clouston as a four on the RRASOR scale. The RRASOR system—pronounced "razor," as in Occam's razor—is based on just the four factors listed below:

1. Prior sexual offenses

None	0
1 conviction or 1–2 charges	1
2–3 convictions or 3–5 charges	2
4+ convictions or 6+ charges	3

2. Age of release (current age)

More than 25	0
Less than 25	1

3. Victim gender

Only females	0
Any males	1

4. Relationship to victim

Only related	0
Any nonrelated	1

SOURCE: John Monahan and Laurens Walker, *Social Science in Law: Cases and Materials* (2006).

Clouston would receive one point for victimizing a male, one for victimizing a nonrelative, and two more because he had three previous sex-offense charges. It's hard to feel any pity for Clouston, but this seventy-one-year-old could be funneled toward lifetime commitment based in part upon crimes for which he'd never been convicted. What's more, this statutory trigger expressly discriminates based on the sex of his victims. These factors are not chosen to assess the relative blame-worthiness of different inmates. They are solely about predicting the likelihood of recidivism. If it turned out that wholly innocent conduct (putting barbecue sauce on ice cream) had a statistically valid, positive correlation with recidivism, the RRASOR system at least in theory would condition points on such behavior.

This Super Crunching cutoff of course doesn't mandate civil commitment; it just mandates that humans consider whether he should be committed as a "sexually violent predator." State officials in exercising this decision not infrequently wave off the Super Crunching prediction. Since the statute was passed, the attorney general's office has sought commitments against only about 70 percent of the inmates who scored a four or more on the risk assessment, and only about 70 percent of the time have courts granted the state's petition to commit these inmates.

The Virginia statute thus channels discretion, but it does not obliterate it. To cede complete decision-making power to lock up a human to a statistical algorithm is in many ways unthinkable. Complete deference to statistical prediction in this or other contexts would almost certainly lead to the odd decision that at times we "know" is going to be wrong. Indeed, Paul Meehl long ago worried about the "case of the broken leg." Imagine that a Super Cruncher is trying to predict whether individuals will go to the movies on a certain night. The Super Crunching formula might predict on the basis of twenty-five statistically validated factors that Professor Brown has an 84 percent probability of going to a movie next Friday night. Now suppose that we also learn that Brown has a compound leg fracture from an accident a few days ago and is immobilized in a hip cast.

Meehl understood that it would be absurd to rely on the actuarial prediction in the face of this new piece of information. By solely relying on the regression or even relegating the expert's opinion to merely being an additional input to this regression, we are likely to make the wrong decision. A statistical procedure cannot estimate the causal impact of rare events (like broken legs) because there simply aren't enough data concerning them to make a credible estimate. The rarity of the event doesn't mean that it will not have a big impact when the event does in fact occur. It just means that statistical formulas will not be able to capture the impact. If we really care about making an accurate prediction in such circumstances, we need to have some kind of discretionary escape hatch—some way for a human to override the prediction of the formula.

The problem is that these discretionary escape hatches have costs too. "People see broken legs everywhere," Snijders says, "even when they are not there." The Mercury astronauts insisted on a literal escape hatch. They balked at the idea of being bolted inside a capsule that could only be opened from the outside. They demanded discretion. However, it was discretion that gave Liberty Bell 7 astronaut Gus Grissom the opportunity to panic upon splashdown. In Tom Wolfe's memorable account, Grissom "screwed the pooch" when he prematurely blew the seventy explosive bolts securing the hatch before the Navy SEALs were able to secure floats. The space capsule sank and Gus nearly drowned.

System builders must carefully consider the costs as well as the benefits of delegating discretion. In context after context, decision makers who wave off the statistical predictions tend to make poorer decisions. The expert override doesn't do worse when a true broken leg event occurs. Still, experts are overconfident in their ability to beat the system. We tend to think that the restraints are useful for the other guy but not for us. So we don't limit our overrides to the clear cases where the formula is wrong; we override where we think we know better. And that's when we get in trouble. Parole and Civil Commitment boards that make exceptions to the statistical algorithm and release inmates who are predicted to have a high probability of violence tend time and again to find that the high probability parolees have higher recidivism rates than those predicted to have a low probability. Indeed, in Virginia only one man out of the dozens civilly committed under the SVPA has ever been subsequently released by a judge who found him—notwithstanding his RRASOR score—no longer to be a risk to society. Once freed, this man abducted and sodomized a child and now is serving a new prison sentence.

There is an important cognitive asymmetry here. Ceding complete control to a statistical formula inevitably will give rise to the intolerable result of making some decisions that reason tells must be wrong. The "broken leg" hypothetical is cute, but unconditional adherence to statistical formulas will lead to powerful examples of tragedy—organs being transplanted into people we know can't use them. These rare but salient anecdotes will loom large in our consciousness. It's harder to keep in mind evidence that discretionary systems, where experts are allowed to override the statistical algorithms, tend to do worse.

What does all this mean for human flourishing? If we care solely about getting the best decisions overall, there are many contexts where we need to relegate experts to mere supporting roles in the decision-making process. We, like the Mercury astronauts, probably can't tolerate a system that forgoes any possibility of human override. At a minimum, however, we should keep track of how experts fare when they wave off the suggestions of the formulas. The broken leg hypothetical teaches us that there will, of course, be unusual circumstances where we'll have good reason for ignoring statistical prediction and going with what our gut and our reason tell us to do. Yet we also need to keep an eye on how often we get it wrong and try to limit our own discretion to places where we do better than machines. "It is critical that the level, type, and circumstances of over-ride usage be monitored on an ongoing basis," University of Massachusetts criminologists James Byrne and April Pattavina wrote recently. "A simple rule of thumb for this type of review is to apply a 10 percent rule: if more than 10 percent of the agency's risk scoring decisions are being changed, then the agency has a problem in this area that needs to be resolved." They want to make sure that override is limited to fairly rare circumstances. I'd propose instead that if more than half of the overrides are getting it wrong, then humans, like the Mercury astronauts, are overriding too much.

This is in many ways a depressing story for the role of flesh-and-blood people in making decisions. It looks like a world where human discretion is sharply constrained, where humans and their decisions are controlled by the output of machines. What, if anything, in the process of prediction can we humans do better than the machines?

What's Left for Us to Do?

In a word, hypothesize. The most important thing that is left to humans is to use our minds and our intuition to guess at what variables should and should not be included in statistical analysis. A statistical regression can tell us the weights to place upon various factors (and simultaneously tell us how precisely it was able to estimate these weights). Humans, however, are crucially needed to generate the hypotheses about what causes what. The regressions can test whether there is a causal effect and estimate the size of the causal impact, but somebody (some body, some human) needs to specify the test itself.

Consider, for example, the case of Aaron Fink. Fink was a California urologist and an outspoken advocate of circumcision. (He was the kind of guy who self-published a book to promote his ideas.) In 1986, the *New England Journal of Medicine* published a letter of his which proposed that uncircumcised men would be more susceptible to HIV infection than circumcised men. At the time, Fink didn't have any data, he just had the idea that the cell of the prepuce (the additional skin on an uncircumcised male) might be susceptible to infection. Fink also noticed that countries like Nigeria and Indonesia where only about 20 percent of men are uncircumcised had a slower spread of AIDS than in countries like Zambia and Thailand where 80 percent of men are uncircumcised. Seeing through the sea of data to recognize a correlation that had eluded everyone else was a stroke of brilliance.

Before Fink died in 1990, he was able to see the first empirical verification of his idea. Bill Cameron, an AIDS researcher in Kenya, hit upon a powerful test of Fink's hypothesis. Cameron and his colleagues found 422 men who visited prostitutes in Nairobi, Kenya, in 1985 (85 percent of these prostitutes were known to be HIV positive) and subsequently went to a clinic for treatment of a non-HIV STD. Like Ted Ruger's Supreme Court study, Cameron's test was prospective. Cameron and his colleagues counseled the men on HIV, STDs, and condom use and asked them to refrain from further prostitute contact. The researchers then followed up with these men on a monthly basis for up to two years, to see if and under what circumstances the men became HIV positive. Put simply, they found that uncircumcised men were 8.2 times more likely to become HIV positive than circumcised men.

This small but powerful study triggered a cascade of dozens of studies confirming the result. In December 2006, the National Institutes of Health stopped two randomized trials that it was running in Kenya and Uganda because it became apparent that circumcision reduced a man's risk of contracting AIDS from heterosexual sex by about 65 percent. Suddenly, the Gates Foundation is considering paying for circumcision in high risk countries.

What started as a urologist's hunch may end up saving hundreds of thousands of lives. Yes, there is still a great role for deductive thought. The Aristotelian approach to knowledge remains important. We still need to theorize about the true nature of things, to speculate. Yet unlike the old days, where theorizing was an end in itself, the Aristotelian approach will increasingly be used at the beginning as an input to statistical testing. Theory or intuition may lead the Finks of the world to speculate that X and Y cause Z. But Super Crunching (by

the Camerons) will then come decisively into play to test and parame-
terize the size of the impact.

The role of theory in *excluding* potential factors is especially important. Without theory or intuition, there is a literal infinity of possible
causes for any effect. How are we to know that what a vintner had for
lunch when he was seven doesn't affect who he might fall in love with
or how quaffable his next vintage might be? With finite amounts of
data, we can only estimate a finite number of causal effects. The
hunches of human beings are still crucial in deciding what to test and
what not to test.

The same is even more true for randomized testing. People have to
figure out in advance what to test. A randomized trial only gives you
information about the causal impact of some treatment versus a control group. Technologies like Offermatica are making it a lot cheaper
to test dozens of separate treatments' effects. Yet there's still a limit to
how many things can be tested. It would probably be just a waste of
money to test whether an ice-cream diet would be a healthy way to
lose weight. But theory tells me that it might not be a bad idea to test
whether financial incentives for weight loss would work.

So the machines still need us. Humans are crucial not only in deciding what to test, but also in collecting and, at times, creating the
data. Radiologists provide important assessments of tissue anomalies
that are then plugged into the statistical formulas. Same goes for
parole officials who subjectively judge the rehabilitative success of
particular inmates. In the new world of database decision making,
these assessments are merely inputs for a formula, and it is statistics,
and not experts, that determine how much weight is placed on the assessments.

Albert Einstein said the "really valuable thing is intuition." In
many ways, he's still right. Increasingly, though, intuition is a precur-
sor to Super Crunching. In case after case, traditional expertise innocent of statistical analysis is losing out to unaided intuition. As Paul
Meehl concluded shortly before he died:

There is no controversy in social science which shows such a large body of qualitatively diverse studies coming out so uniformly in the same direction as this one. When you are pushing over 100 investigations, predicting everything from the outcome of football games to the diagnosis of liver disease, and when you can hardly come up with a half dozen studies showing even a weak tendency in favor of the clinician, it is time to draw a practical conclusion.

It is much easier to accept these results when they apply to someone else. Few people are willing to accept that a crude statistical algorithm based on just a handful of factors could outperform them. Universities are loath to accept that a computer could select better students. Book publishers would be loath to delegate the final say in acquiring manuscripts to an algorithm.

At some point, however, we should start admitting that the superiority of Super Crunching is not just about the other guy. It's not just about baseball scouts and wine critics and radiologists and . . . the list goes on and on. Indeed, by now I hope to have convinced you that something real is going on out there. Super Crunching is impacting real-world decisions in many different contexts that touch us as consumers, as patients, as workers, and as citizens.

Kenneth Hammond, the former director of Colorado's Center for Research on Judgment and Policy, reflects with some amusement on the resistance of clinical psychologists to Meehl's overwhelming evidence:

One might ask why clinical psychologists are offended by the discovery that their intuitive judgments and predictions are (almost) as good as, but (almost) never better than, a rule. We do not feel offended at learning that our excellent visual perception can often be improved in certain circumstances by the use of a tool (e.g., rangefinders, telescopes, microscopes). The answer seems to be that tools are used by clerks (i.e., someone without professional training); if psychologists are no different, then that demeans the status of the psychologist.

This transformation of clinicians to clerks is indicative of a larger trend. Something has happened out there that has triggered a shift of discretion from traditional experts to a new breed of Super Crunchers, the people who control the statistical equations.

CHAPTER 6

Why Now?

My nephew Marty has a T-shirt that says on the front: "There are 10 types of people in the world..." If you read these words and are trying to think of what the ten types are, you've already typed yourself.

The back of the shirt reads: "Those that understand binary, and those that don't." You see, in a digitalized world, all numbers are represented by 0s and 1s, so what we know as the number 2 is represented to a computer as 10. It's the shift to binary bytes that is at the heart of the Super Crunching revolution.

More and more information is digitalized in binary bytes. Mail is now email. From health care files to real estate and legal filings, electronic records are everywhere. Instead of starting with paper records and then manually inputting information, data are increasingly captured electronically in the very first instance—such as when we swipe our credit card or the grocery clerks scan our deodorant purchase at the checkout line. An electronic record of most consumer purchases now exists.

And even when the information begins on paper, inexpensive scanning technologies are unlocking the wisdom of the distant and not-so-distant past. My brother-in-law used to tell me, "You can't Google dead trees." He meant that it was impossible to search the text of books. Yet now you can. For a small monthly fee, Questia.com will give you full text access to over 67,000 books. Amazon.com's "Search Inside the Book" feature allows Internet users to read snippets from full text searches of over 100,000 volumes. And Google is attempting something that parallels the Human Genome Project in both its scope and scale. The Human Genome Project had the audacity to think that within the space of thirteen years it could sequence three billion genes. Google's "Book Search" is ambitiously attempting to scan the full text of more than thirty million books in the next ten years. Google intends to scan every book ever published.

From 90 to 3,000,000

The increase in accessibility to digitalized data has been a part of my own life. Way back in 1989 when I had just started teaching, I sent six testers out to Chicagoland new car dealerships to see if dealers discriminated against women or minorities. I trained the testers to follow a uniform script that told them how to bargain for a car. The testers even had uniform answers to any questions the salesman might ask (including "I'm sorry but I'm not comfortable answering that"). The testers walked alike, talked alike. They were similar on every dimension I could think of except for their race and sex. Half the testers were white men and half were either women or African-Americans. Just like a classic fair housing test, I wanted to see if women or minorities were treated differently than white men.

They were. White women had to pay 40 percent higher markups

than white men; black men had to pay more than twice the markup, and black women had to pay more than three times the markup of white male testers. My testers were systematically steered to salespeople of their own race and gender (who then gave them worse deals).

The study got a lot of press when it was published in the *Harvard Law Review*. *Primetime Live* filmed three different episodes testing whether women and minorities were treated equally not just at car dealerships but at a variety of retail establishments. A lot of people were disturbed by film clips of shoe clerks who forced black customers to wait and wait for service even though no one else was in the store. More importantly, the study played a small role in pushing the retail industry toward no-haggle purchasing.

A few years after my study, Saturn decided to air a television commercial that was centrally about Saturn's unwillingness to discriminate. The commercial was composed entirely of a series of black-and-white photographs. In a voiceover narrative, an African-American man recalls his father returning home after purchasing a car and feeling that he had been mistreated by the salesman. The narrator then says maybe that's why he feels good about having become a salesperson for Saturn. The commercial is a remarkable piece of rhetoric. The stark photographic images are devoid of the smiles that normally populate car advertisements. Instead there is a heartrending shot of a child realizing that his father has been mistreated because of their shared race and the somber but firmly proud shot of the grown, grim-faced man now having taken on the role of a salesman who does not discriminate. The commercial does not explicitly mention race or Saturn's no-haggle policy—but few viewers would fail to understand that race was a central cause of the father's mistreatment.

The really important point is that all this began with six testers bargaining at just ninety dealerships. While I ultimately did a series of follow-up studies analyzing the results of hundreds of additional bargains, the initial uproar came from a very small study. Why so small? It's hard to remember, but this was back in the day before the Internet. Laptop computers barely existed and were expensive and bulky. As a result, all of my data were first collected on paper and then had to be hand-entered (and re-entered) into computer files for analysis. Technologically, back then, it was harder to create digital data.

Fast-forward to the new millennium, and you'll still find me crunching numbers on race and cars. Now, however, the datasets are much, much bigger. In the last five years, I've helped to crunch numbers in massive class-action litigation against virtually all of the major automotive lenders. With the yeoman help of Vanderbilt economist Mark Cohen (who really bore the laboring oar), I have crunched data on more than three million car sales.

While most consumers now know that the sales price of a car can be negotiated, many do not know that auto lenders, such as Ford Motor Credit or GMAC, often give dealers the option of marking up a borrower's interest rate. When a car buyer works with the dealer to arrange financing, the dealer normally sends the customer's credit information to a potential lender. The lender then responds with a private message to the dealer that offers a "buy rate"—the interest rate at which the lender is willing to lend. Lenders often will pay a dealer—sometimes thousands of dollars—if the dealer can get the consumer to sign a loan with an inflated interest rate. For example, Ford Motor Credit tells a dealer that it was willing to lend Susan money at a 6 percent interest rate, but that they would pay the dealership $2,800 if the dealership could get Susan to sign an 11 percent loan. The borrower would never be told that the dealership was marking up the loan. The dealer and the lender would then split the expected profits from the markup, with the dealership taking the lion's share.

In a series of cases that I worked on, African-American borrowers challenged the lenders' markup policies because they disproportionately harmed minorities. Cohen and I found that on average white borrowers paid what amounted to about a $300 markup on their loans, while black borrowers paid almost $700 in markup profits. Moreover, the distribution of markups was highly skewed. Over half of white borrowers paid no markup at all, because they qualified for loans where markups were not allowed. Yet 10 percent of GMAC borrowers paid more than $1,000 in markups and 10 percent of the Nissan customers paid more than a $1,600 markup. These high markup borrowers were disproportionately black. African-Americans were only 8.5 percent of GMAC borrowers, but paid 19.9 percent of the markup profits. The markup difference wasn't about credit scores or default risk; minority borrowers with good credit routinely had to pay higher markups than white borrowers with similar credit scores.

These studies were only possible because lenders now keep detailed electronic records of every transaction. The one variable they don't keep track of is the borrower's race. Once again, though, technology came to the rescue. Fourteen states (including California) will, for a fee, make public the information from their driver's license database—information that includes the name, race, and Social Security number of the driver. Since the lenders' datasets also included the Social Security numbers of their borrowers (so that they could run credit checks), it was child's play to combine the two different datasets. In fact, because so many people move around from state to state, Cohen and I were able to identify the race of borrowers for thousands upon thousands of loans that took place in all fifty states. We'd know the race of a lot of people who bought cars in Kansas, because sometime earlier or later in their lives they took out a driver's license in California. A study that would have been virtually impossible to do ten years earlier had now become, if not easy, at least relatively straightforward. And in fact, Cohen and I did the study over and over as the cases against all the major automotive lenders moved forward. The cases have been a resounding success: lender after lender has agreed to cap the amount that dealerships can mark up loans. All borrowers regardless of their race are now protected by caps that they don't even know about. Unlike my initial test of a few hundred negotiations, these statistical studies of millions of transactions were only possible because the information now is stored in readily accessible digital records.

Trading in Data

The willingness of states to sell information on the race of their own citizens is just a small part of the commercialization of data. Digitalized data has become a commodity. And both public and private vendors have found the value of aggregating information. For-profit database aggregators like Acxiom and ChoicePoint have flourished. Since its founding in 1997, ChoicePoint has acquired more than seventy smaller database companies. It will sell clients one file that contains not only your credit report but also your motor-vehicle, police, and property records together with birth and death certificates and marriage and divorce decrees. While much of this information was already publicly available, ChoicePoint's billion dollars in annual revenue suggests that there's real value in providing one-stop data-shopping.

And Acxiom is even larger. It maintains consumer information on nearly every household in the United States. Acxiom, which has been called "one of the biggest companies you've never heard of," manages twenty billion customer records (more than 850 terabytes of raw data—enough to fill a 2,000-mile tower of one billion diskettes).

Like ChoicePoint, a lot of Acxiom's information is culled from public records. Yet Acxiom combines public census data and tax records with information supplied by corporations and credit card companies that are Acxiom clients. It is the world's leader in CDI, consumer data integration. In the end, Acxiom probably knows the catalogs you get, what shoes you wear, maybe even whether you like dogs or cats. Acxiom assigns every person a thirteen-digit code and places them in one of seventy "lifestyle" segments ranging from "Rolling Stones" to "Timeless Elders." To Acxiom, a "Shooting Star" is someone who is thirty-six to forty-five, married, no kids yet, wakes up early and goes for runs, watches *Seinfeld* reruns, and travels abroad. These segments are so specific to the time of life and triggering events (such as getting married) that nearly one-third of Americans change their segment each year. By mining its humongous database, Acxiom not only knows what segment you are in today but it can predict what segment you are likely to be in next year.

The rise of Acxiom shows how commercialization has increased the fluidity of information across organizations. Some large retailers like Amazon.com and Wal-Mart simply sell aggregate customer transaction information. Want to know how well Crest toothpaste sells if it's placed higher on the shelf? Target will sell you the answer. But Acxiom also allows vendors to trade information. By providing Acxiom's transaction information about its individual customers, a retailer can gain access to a data warehouse of staggering proportions.

Do the Mash

The popular Internet mantra "information wants to be free" is centrally about the ease of liberating digital data so that it can be exploited by multiple users. The rise of database decision making is driven by the increasing access to what was OPI—other people's information. Until recently, many datasets—even inside the same corporation—couldn't easily be linked together. Even a firm that maintained two different datasets often had trouble linking them if the datasets had incompatible formats or were developed by different software companies. A lot of data were kept in isolated "data silos."

These technological compatibility constraints are now in retreat. Data files in one format are easily imported and exported to other formats. Tagging systems allow single variables to have multiple names. So a retailer's extra-large clothes can simultaneously be referred to as "XL" and "TG" (for the French term *très grande*). Almost gone are the days where it was impossible to link data stored in non-compatible proprietary formats.

What's more, there is a wealth of non-proprietary information on the web just waiting to be harvested and merged into pre-existing datasets. "Data scraping" is the now-common practice of programming a computer to surf to a set of sites and then systematically copy information into a database. Some data scraping is pernicious—such as when spammers scrape email addresses off websites to create their spam lists. But many sites are happy to have their data taken and used by others. Investors looking for fraud or accounting hijinks can scrape data from quarterly SEC filings of all traded corporations. I've used computer programs to scrape data for eBay auctions to create a dataset for a study I'm doing about how people bid on baseball cards.

A variety of programmers have combined the free geographical information from Google maps with virtually any other dataset that contains address information. These data "mashups" provide striking visual maps that can represent the crime hot spots or campaign contributions or racial composition or just about anything else. Zillow.com mashes up public tax information about house size and other neighborhood characteristics together with information about recent neighborhood sales to produce beautiful maps with predicted housing values.

A "data commons" movement has created websites for people to post and link their data with others. In the last ten years, the norm of sharing datasets has become increasingly grounded in academics. The premier economics journal in the United States, the *American Economic Review*, requires that researchers post to a centralized website all the data backing up their empirical articles. So many researchers are posting their datasets to their personal web pages that it is now more likely than not that you can download the data for just about any empirical article by just typing a few words into Google. (You can find tons of my datasets at www.law.yale.edu/ayres/.)

Data aggregators like Acxiom and ChoicePoint have made an art of finding publicly available information and merging it into their pre-existing databases. The FBI has information on car theft in each city for each year; Allstate has information about how many anti-theft devices were used in particular cities in particular years. Nowadays, regardless of the initial digital format, it has become a fairly trivial task to link these two types of information together. Today, it's even possible to merge datasets when there doesn't exist a single unique identifier, such as the Social Security number, to match up observations. Indirect matches can be made by looking for similar patterns. For example, if you want to match house purchases from two different records, you might look for purchases that happened on the same day in the same city.

Yet the art of indirect matching can also be prone to error. Database Technologies (DBT), a company that was ultimately purchased by ChoicePoint, got in a lot of trouble for indirectly identifying felons before the 2000 Florida elections. The state of Florida hired DBT to create a list of potential people to remove from the list of registered voters. DBT matched the database of registered voters to lists of convicted felons not just from Florida but from every state in the union. The most direct and conservative means to match would have been to use the voter's name and date of birth as necessary identifiers. But DBT, possibly under direction from Florida's Division of Elections, cast a much broader net in trying to identify potential convicts. Its matching algorithm required only a 90 percent match between the name of the registered voter and the name of the convict. In practice this meant that there were lots of false positives, registered voters who were wrongly identified as possibly being convicts. For example, the Rev. Willie D. Whiting, Jr., a registered voter from Tallahassee, was initially told that he could not vote because someone named Willie J. Whiting, born two days later, had a felony conviction. The Division of Elections also required DBT to perform "nickname matches" for first names and to match on first and last names regardless of their order— so that the name Deborah Ann would also match the name Ann Deborah, for example.

The combination of these low matching requirements together with the broad universe of all state felonies produced a staggeringly large list of 57,746 registered Floridians who were identified as convicted felons. The concern was not just with the likely large number of false positives, but also with the likelihood that a disproportionate number of the so-called purged registrations would be for African-American voters. This is especially true because the algorithm was not relaxed when it came to race. Only registered voters who exactly matched the race of the convict were subject to exclusion from the voting rolls. So while Rev. Whiting, notwithstanding a different middle initial and birth date, could match convict Willie J. Whiting, a white voter with the same name and birth date would not qualify because the convict Whiting was black.

Mashups and mergers of datasets are easier today than ever before. But the DBT convict list is a cautionary tale. The new merging technology can fail through either inadvertent or advertent error. As the size of datasets balloons almost beyond the scope of our imagination, it becomes all the more important to continually audit them to check for the possibility of error. What makes the DBT story so troubling is that the convict/voter data seemed so poorly matched relative to the standards of modern-day merging and mashing.

Technology or Techniques?

The technological advances in the ability of firms to digitally capture and merge information have helped trigger the commodification of data. You're more likely to want to pay for data if you can easily merge them into your pre-existing database. And you're more likely to capture information if you think someone else is later going to pay you for it. So the ability of firms to more easily capture and merge information helps answer the "Why Now?" question.

At heart, the recent onslaught of Super Crunching is more a story of advances in technology, not statistical techniques. It is not a story about new breakthroughs in the statistical art of prediction. The basic statistical techniques have existed for decades—even centuries. The randomized trials that are being used so devastatingly by companies like Offermatica have been known and used in medicine for years. Over the last fifty years, econometrics and statistical theory have improved, but the core regression and randomization techniques have been available for a long time.

What's more, the timing of the Super Crunching revolution isn't dominantly about the exponential increase in computational capacity. The increase in computer speed has helped, but the increase in computer speed came substantially before the rise of data-based decision making. In the old days, say, before the 1980s, CPUs—"central processing units"—were a real constraint. The number of mathematical operations necessary to calculate a regression goes up exponentially with the number of variables—so if you double the number of controls, you roughly quadruple the number of operations needed to estimate a regression equation. In the 1940s, the Harvard Computation Laboratory employed scores of secretaries armed with mechanical calculators who manually crunched the numbers behind individual regressions. When I was in grad school at MIT in the 1980s, CPUs were so scarce that grad students were only allotted time in the wee hours of the morning to run our programs.

But thanks to Moore's Law—the phenomenon that processor power doubles every two years—Super Crunching has not been seriously hampered by a lack of cheap CPUs. For at least twenty years, computers have had the computation power to estimate some serious regression equations.

The timing of the current rise of Super Crunching has been impacted more by the increase in storage capacity. We are moving toward a world without delete buttons. Moore's Law is better known, but it is Kryder's Law—a regularity first proposed by the chief technology officer for hard drive manufacturer Seagate Technology, Mark Kryder—that is more responsible for the current onslaught of Super Crunching. Kryder successfully noticed that the storage capacity of hard drives has been doubling every two years.

Since the introduction of the disk drive in 1956, the density of information that can be recorded into the space of about a square inch has swelled an amazing 100-million fold. Anyone over thirty remembers the days when we had to worry frequently about filling up our hard disks. Today, the possibility of cheap data storage has revolutionized the possibility for keeping massively large datasets.

And as the density of storage has increased, the price of storage has dropped. Thirty to forty percent annual price declines in the cost per gigabyte of storage continue apace. Yahoo! currently records over twelve terabytes of data daily. On the one hand, this is a massive amount of information—it's roughly equivalent to more than half the information contained in all the books in the Library of Congress. On the other hand, this amount of disk storage does not require acres of servers or billions of dollars. In fact, right now you could add a terabyte of hard drive to your desktop for about $400. And industry experts predict that in a couple of years that price will drop in half.

The cutthroat competition to produce these humongous hard drives for personal consumer products is driven by video. TiVo and other digital video recorders can only remake the world of home video entertainment if they have adequate storage space. A terabyte drive will only hold about eight hours of HDTV (or nearly 14,000 music albums), but you can jam onto it about sixty-six million pages of text or numbers.

Both the compactness and cheapness of storage are important for the proliferation of data. Suddenly, it's feasible for Hertz or UPS to give each employee a handheld machine to capture and store individual transaction data that are only periodically downloaded to a server. Suddenly, every car includes a flash memory drive, a mini-black box recorder to tell what was happening at the time of an accident.

The abundance of supercheap storage from the tiny flash drives (hidden inside everything from iPods and movie cameras to swimming goggles and birthday cards) to the terabyte server farms at Google and flickr.com have opened up new vistas of data-mining possibilities. The recent onslaught of Super Crunching is dominantly driven by the same technological revolutions that have been reshaping so many other parts of our lives. The timing is best explained by the digital breakthroughs that make it cheaper to capture, to merge, and to store huge electronic databases. Now that mountains of data exist (on hard disks) to be mined, a new generation of empiricists is emerging to crunch it.

Can a Computer Be Taught to Think Like You?

There is, though, one new statistical technique that is an important contributor to the Super Crunching revolution: the "neural network." Predictions using neural network equations are a newfangled competitor to the tried-and-true regression formula. The first neural networks were developed by academics to simulate the learning processes of the human brain. There's a great irony here: the last chapter detailed scores of studies showing why the human brain does a bad job of predicting. Neural networks, however, are attempts to make computers process information like human neurons. The human brain is a network of interconnected neurons that act as informational switches. Depending on the way the neuron switches are set, when a particular neuron receives an impulse, it may or may not send an impulse on to a subsequent set of neurons. Thinking is the result of particular flows of impulses through the network of neuron switches. When we learn from some experience, our neuron switches are being reprogrammed to respond differently to different types of information. When a curious young child reaches out and touches a hot stove, her neuron switches are going to be reprogrammed to fire differently so the next time the hot stove will not look so enticing.

The idea behind computer neural networks is essentially the same: computers can be programmed to update their responses based on new or different information. In a computer, a mathematical "neural network" is a series of interconnected switches that, like neurons, receive, evaluate, and transmit information. Each switch is a mathematical equation that takes and weighs multiple types of input information. If the weighted sum of the inputs in the equation is sufficiently large, the switch is turned on and is sent as informational input for subsequent neural equation switches. At the end of the network is a final switch that collects information from previous neural switches and produces as its output the neural network's prediction. Unlike the regression approach, which estimates the weights to apply to a single equation, the neural approach uses a system of equations represented by a series of interconnected switches.

Just as experience trains our brain's neuron switches when to fire and when not to, computers use historical data to train the equation switches to come up with optimal weights. For example, researchers at the University of Arizona constructed a neural network to forecast winners in greyhound dog racing at the Tucson Greyhound Park. They fed in more than fifty pieces of information from thousands of daily racing sheets—things like the dogs' physical attributes, the dogs' trainers, and, of course, how the dogs did in particular races under particular conditions. Like the haphazard predictions of the curious young child, the weights on these greyhound racing equations were initially set randomly. The neural estimation process then tried out alternative weights on the same historic data over and over again—sometimes literally millions of times—to see which weights for the interconnecting equations produced the most accurate estimates. The researchers then used the weights from this training to predict the outcome of a hundred future dog races.

The researchers even set up a contest between their predictions and three expert habitués of the racetrack. For the test races, the neural network and the experts were each instructed to place $1 bets on a hundred different dogs. Not only did the neural network better predict the winners, but (more importantly) the network's predictions yielded substantially higher payoffs. In fact, while none of the three experts generated positive payoffs with their predictions—the best still lost $60—the neural network won $125. It won't surprise you to learn that lots of other bettors are now relying on neural prediction (if you google neural network and betting, you'll get tons of hits).

You might be wondering what's really new about this technique. After all, plain-old regression analysis also involves using historical data to predict results. What sets the neural network methodology apart is its flexibility and nuance. With traditional regressions, the Super Cruncher needs to specify the specific form of the equation. For example, it's the Super Cruncher who has to tell the machine whether or not the dog's previous win percentage needs to be multiplied by the dog's average place in a race in order to produce a more powerful prediction.

With neural networks, the researcher just needs to feed in the raw information, and the network, by searching over the massively interconnected set of equations, will let the data pick out the best functional form. We don't have to figure out in advance how dogs' different physical attributes interact to make them better racers; we can let the neural training tell us. The Super Cruncher, under either the regression or neural methods, still needs to specify the raw inputs for the prediction. The neural method, however, allows much more fluid estimates of the nature of the impact. As the size of the datasets has increased, it has become possible to allow neural networks to estimate many, many more parameters than have traditionally been accommodated by traditional regression.

But the neural network is not a panacea. The subtle interplay of its weighting schemes is also one of its biggest drawbacks. Because a single input can influence multiple intermediate switches that in turn impact the final prediction, it often becomes impossible to figure out how an individual input is affecting the predicted outcome.

Part and parcel of not knowing the size of the individual influences is not knowing the precision of the neural weighting scheme. Remember, the regression not only tells you how much each input impacts the prediction, it also tells you how accurately it was able to estimate the impact. Thus, in the greyhound example, a regression equation might not only tell you that the dog's past win percentage should be given a weight of .47 but it would also tell you its level of confidence in that prediction: "There's a 95 percent chance that the true weight is between .35 and .59." The neural network, in contrast, doesn't tell you the confidence intervals. So while the neural technique can yield powerful predictions, it does a poorer job of telling you why it is working or how much confidence it has in its prediction.

The multiplicity of estimated weighting parameters (which can often be three times greater with neural networks than with regression prediction) can also lead toward "overfitting" of the training data.* If you have the network "train" itself about the best 100 weights to use on 100 pieces of historical data, the network will be able to precisely predict all 100 outcomes. But exactly fitting the past doesn't guarantee that the neural weights will be good at predicting future outcomes. Indeed, the effort to exactly fit the past with a proliferation of arbitrary weights can actually hinder the ability of neural networks to predict the future. Neural Super Crunchers are now intentionally limiting the number of parameters that they estimate and the amount of time they let the network train to try to reduce this overfitting problem.

"We Shoot Turkeys"

To be honest, the neural prediction methods are sufficiently new that there's still a lot of art involved in figuring out how best to estimate neural predictions. It's not yet clear how far neural prediction will go in replacing regression prediction as the dominant methodology. It is clear, however, that there are real-world contexts in which neural predictions are at least holding their own with regression prediction as far as accuracy. There are even some cases where they outperform traditional regressions.

Neural predictions are even starting to influence Hollywood. Just as Orley Ashenfelter predicted the price of Bordeaux vintages before they had even been tasted, a lawyer named Dick Copaken has had the audacity to think that he can figure out how much a movie will gross before a single frame is even shot. Copaken is a fellow Kansas Citian

---

* "Overfitting" refers to fitting a statistical model that contains too many parameters.

who graduated from Harvard Law School and went on to a very distinguished career as a partner in the Washington office of Covington & Burling. In the past, he's crunched numbers for his legal clients. Years ago he commissioned Lou Harris to collect information on perceptions of car bumper damage. The statistical finding that most people couldn't even see small dents in their bumpers convinced the Department of Transportation to allow manufacturers to use cheaper bumpers that would sustain at most imperceptible dents.

Nowadays, Copaken is using a neural network to crunch numbers for a very different kind of client. After retiring from the practice of law, Dick Copaken founded a company that he named Epagogix (from the Aristotelian idea of inductive learning). The company has "trained" a neural network to try to predict a movie's receipts based primarily on characteristics of the script. Epagogix has been working behind the scenes because most of its clients don't want the world to know what it's doing.

But in a 2006 *New Yorker* article, Malcolm Gladwell broke the story. Gladwell first learned of Epagogix when he was giving a speech to the head honchos of a major film studio. Copaken told me it was the kind of "retreat where they essentially confiscate everybody's BlackBerry and cell phone, move them to some off-campus site and for a few days they try to think the great thoughts. . . . And they usually invite some guru of the moment to come and speak with them and help them as they sort through their thinking. And this particular year it was Malcolm Gladwell." Even though Gladwell's job was to tell stories to the executives, he turned the tables and asked them to tell him about some idea that was going to reshape the way films are made and viewed in the next century. "The chairman of the board started to tell him," Copaken said, "about . . . this company that does these neural network projections and they are really amazingly accurate. And then, although this was all supposed to be fairly hush-hush, . . . the head of the studio chimed in and began to give some specifics about just how accurate we were in a test that we had done for the studio."

The studio head was bragging about the results of a paradigm-shifting experiment in which Epagogix was asked to predict the gross revenues of nine motion pictures just based on their scripts—before the stars or the directors had even been chosen. What made the CEO so excited was that the neural equations had been able to accurately predict the profitability of six out of nine films. On a number of the films, the formula's revenue prediction was within a few million dollars of the actual gross.

Six out of nine isn't perfect, but traditionally studios are only accurate on about a third of their predictions of gross revenues. When I spoke with Copaken, he was not shy about putting dollars to this difference. "For the larger studios, if they both had the benefit of our advice and the discipline to adhere to it," he said, "they could probably net about a billion dollars or more per studio per year." Studios are effectively leaving a billion dollars a year on the ground.

Several studios have quickly (but quietly) glommed on to Epagogix's services. The studios are using the predictions to figure out whether it's worth spending millions of dollars to make a movie. In the old days, the phrase "to shoot a turkey" meant to make a bad picture. When someone at Epagogix says, "We shoot turkeys," it means just the opposite. They prevent bad pictures from ever coming into existence.

Epagogix's neural equations have also let studios figure out how to improve the expected gross of a film. The formula not only tells you what to change but tells you how much more revenue the change is likely to bring in. "One time they gave us a script that just had too many production sites," Copaken said. "The model told me the audience was going to be losing its place. By moving the action to a single city, we predicted that they would increase revenues and save on production costs."

Epagogix is now working with an outfit that produces about three to four independent films a year with budgets in the $35–50 million range. Instead of just reacting to completed scripts, Epagogix will be helping from the get-go. "They want to work with us in a collegial, collaborative fashion," Copaken explained, "where we will work directly with their writers ... in developing the script to optimize the box office."

But studios that want to maximize profits also have to stop paying stars so much money. One of the biggest neural surprises is that most movies would make just as much money with less established (and therefore less expensive) actors. "We do take actors and directors into account," Copaken says. "It turns out to be a surprisingly small factor in terms of its overall weighting in the box office results." It matters a lot *where* the movie is set. But the name of the stars or directors, not so much. "If you look at the list of the 200 all-time best-grossing movies," Copaken says, "you will be shocked at how few of them have actors who were stars at the time those films were released."

Old habits, like the star system, die hard. Copaken says that studio execs "still aren't listening" when it comes to cutting the ad budget or substituting lesser-named actors. The neural equation says that stars and ads often aren't worth what they cost. Copaken points out, "Nobody really knew about Harrison Ford until after *Star Wars*."

Epagogix isn't on any kind of crusade to hurt stars. In fact, the powerful Endeavor Agency is interested in using Epagogix's services for its own clients. Copaken recently spent a morning with one of Endeavor's founders, the indomitable Ari Emmanuel. Ari Emmanuel is apparently the inspiration for the character Ari Gold, in the HBO series *Entourage*. "He drove me to a couple of major studios to meet with the head of Paramount and the head of Universal," Copaken said. "En route, he must have fielded seventy phone calls, from everybody from Sacha Baron Cohen to Mark Wahlberg to an agent for Will Smith." Endeavor thinks that Epagogix can not only help its clients decide whether to agree to act in a movie, but it could also help them decide whether to get their money up front or roll the dice on participating in back-end profits. In an Epagogix world, some stars may ultimately be paid less, but the savvy stars will know better how to structure their contracts.

It shouldn't surprise you, however, that many players in the industry are not receptive to the idea of neural prediction. Some studios are utterly closed-minded to the idea that statistics could help them decide whether to greenlight a project. Copaken tells the extraordinary story of bringing two hedge fund managers to meet with a studio head. "These hedge fund guys had raised billions of dollars," Copaken explained, "and they were prepared to start with $500 million to fund films that would pass muster by our test and be optimized for box office. And obviously if the thing worked, they were prepared to expand rapidly. So there was a huge source of money on the table there." Copaken thought that he could at least pique the studio's interest with all of that outside money.

"But the meeting was not going particularly well and there was just a lot of resistance to this new way of thinking," Copaken said. "And finally one of these hedge fund guys sort of jumped into the discourse and said, 'Well, let me ask you a question. If Dick's system here gets it right fifty times out of fifty times, are you telling me that you wouldn't take that into account to change the way you decide which movies to make or how to make them?' And the guy said, 'No, that's absolutely right. We would not even if he were right fifty times out of fifty times. . . . [S]o what if we are leaving a billion dollars of the shareholders' money on the table; that is shareholders' money. . . . Whereas if we change the way we do this, we might antagonize various people. We might not be invited. Our wives wouldn't be invited to the parties. People would get pissed at us. So why mess with a good thing?' "

Copaken was completely depressed when he walked out of the meeting, but when he looked over he noticed that the hedge fund guys were grinning from ear to ear. He asked them why they were so happy. They told him, "You don't understand, Dick. We make our fortunes by identifying small imperfections in the marketplace. They are usually tiny and they are usually very fleeting and they are immediately filled by the efficiency of the marketplace. But if we can discover these things and we can throw massive resources at these opportunities, fleeting and small though they may be, we end up making lots of money before the efficiency of the marketplace closes out that opportunity. What you just showed us here in Hollywood is a ten-lane paved highway of opportunity. It's like they are committed to doing things the wrong way and there seems to be so much energy in the culture and commitment to doing it the wrong way, it creates a fantastic opportunity for us that is much more durable and enduring than anything we've ever seen."

The very resistance of some studios creates more of an opportunity for outsiders to come in and see if optimized scripts really do sell more tickets. Epagogix itself is putting its money where its mouth is. Copaken is planning to remake a movie that was a huge commercial disappointment. With the help of the neural network, he thinks just a few simple changes to the script could generate a twenty-three fold increase in the gross. Copaken's lined up a writer and plans to commission a script to make just these changes. We may soon see whether a D.C. lawyer armed with reams of data can perform a kind of cinematographic alchemy.

The screenwriter William Goldman famously claimed that when it comes to picking movies, "Nobody, *nobody*—not now, not ever—knows the least goddamn thing about what is or isn't going to work at the box office." And maybe nobody does. Studio execs, even after years of trial and error, have trouble putting the right weights on the components of a story. Unlike machines, they can emotionally experience a story, but this emotion is a double-edged sword. The relative success of Epagogix's equations stem, in part, from its dispassionate weighting of what works.

Why Not Now?

Teasing out the development of technology and techniques helps explain why the Super Crunching revolution didn't happen earlier. Still, we should also ask the inverse question: why are some industries taking so long to catch the wave? Why have some decisions been resistant to data-driven thinking?

Sometimes the absence of Super Crunching isn't a problem of foot-dragging or unreasonable resistance. There are loads of decisions about which there just isn't adequate historical data to do any kind of statistical test, much less a Super Crunch. Should Google buy YouTube? This kind of one-off question is not readily amenable to data-driven thinking. Super Crunching requires analysis of the results of repeated decisions. And even when there are repeated examples, it's sometimes hard to quantify success. Law schools must decide every year which applicants to admit. We have lots of information about the applicants, and tons of data about past admitted students and the careers they've gone on to have. But what does it mean to be successful after graduation? The most obvious proxy, salary, isn't a great indicator; a leader in government or public interest law might have a relatively low salary, but still make us proud. If you can't measure what you're trying to maximize, you're not going to be able to rely on data-driven decisions.

Nonetheless, there are still many areas where metrics of success and plentiful historical data are just waiting to be mined. While data-driven thinking has been on the rise throughout society, there are still plenty of pockets of resistance that are ripe for change.

There's almost an iron-clad law that it's easier for people to warm up to applications of Super Crunching outside of their own area of expertise. It's devilishly hard for traditional, non-empirical evaluators to even consider the possibility that quantified predictions might do a better job than they can do on their own home turf. I don't think this is primarily because of blatant self-interest in trying to keep our jobs. We humans just overestimate our ability to make good decisions and we're skeptical that a formula that necessarily ignores innumerable pieces of information could do a better job than we could.

So let's turn the light onto the process of publishing books itself. Couldn't Super Crunching help Bantam or its parent, Random House, Inc., decide what to publish? Of course not. Book publishing is too much of an art to be susceptible to Super Crunching. But let's start small. Remember, I already showed how randomized trials helped test titles for this book. Why can't a regression help choose at least the title of books? Turns out Lulu.com has already run this regression. They estimated a regression equation to help predict whether a book's title is going to be a best-seller.

Atai Winkler, a British statistician, created a dataset on the sales of every novel to top the *New York Times* Bestseller List from 1955 to 2004 together with a control group of less successful novels by the same authors. With more than 700 titles, he then estimated a regression to predict the likelihood of becoming a best-seller. The regression tested for the impact of eleven different characteristics (Is the title in the form "The _____ of _____"? Does the title include the name of a person or place? Does it begin with a verb?).

It turns out that figurative titles are more likely to produce best-sellers than are literal ones. It also matters whether the first word of a title is a verb, pronoun, or exclamation. And, contrary to publishing wisdom, shorter isn't necessarily better: a title's length does not significantly affect book sales. All told, the regression produced predictions that were much better than random guesses. "It guessed right in nearly 70 percent of cases," Winkler said. "Given the nature of the data and the way tastes change, this is very good." But Winkler didn't want to over-claim. "Whether a book gets to the best-seller list," he said, "depends a lot on the other books that happen to be there that week on the list. Only one of them could be the best-seller."

The results aren't perfect. While Agatha Christie's *Sleeping Murder* claimed the top spot among all of the titles Winkler analyzed, the model predicted that *The Da Vinci Code* had only a 36 percent chance of becoming a best-seller.

Even with its flaws, this is a web application that's both fun and a bit addictive. Just type in your proposed title at Lulu.com/titlescorer and bam, the applet gives you a prediction of success for any title you might imagine. You can even use the "Titlefight" feature to pit two competing title ideas against each other. Of course, this isn't really a test of whether your book will become a best-seller. It is a test of whether the title of someone like Jane Smiley will take off or not. Yet even if you've never had a book at the top of the best-sellers list, wouldn't you want to know how your title scored? (I did. Even though the book is nonfiction, *Super Crunchers* predicted a 56.8 percent chance of success. From Lulu's lips to God's ears.)

But why stop at the title of the book? Why not crunch the content?

My first reaction is again, nah, that would never work. It's impossible to code whether a book is well written. But this might just be the iron law of resistance talking. Beware of the person who says, "You could never quantify what I do."

If Epagogix's analysis of plots can predict movie sales, why couldn't an analysis of plots help predict novel sales? Indeed, novels should be even easier to code because you don't have the confounding influences of temperamental actors and the possibility of botched or beautiful cinematography. The text is all there is. You might even be able to start by coding the very same criteria that Epagogix uses for movie scripts. The economic criteria for success also exist in abundance. Nielsen BookScan provides its subscribers with weekly point-of-sale data on how well books are selling at most major book retailers. So there are tons of data on sales success just waiting to be crunched. Instead of crudely predicting the probability of whether you hit the top of the best-seller list or not, you could try to predict the total sales based on a lot more than just the title.

Yet no one in publishing is rushing to be the first on the block to publicly use number crunching to choose what books to buy or how to make them better. A large part of me viscerally resists the idea that a nonfiction book could be coded or that Super Crunching could improve the content of this book. But another part of me has in fact already data mined a bit on what makes for success in nonfiction publishing.

As a law professor, my primary publishing job is to write law review articles. I don't get paid for them, but a central measure of an article's success is the number of times the articles have been cited by other professors. So with the help of a full-time number-crunching assistant named Fred Vars, I went out and analyzed what caused a law review article to be cited more or less. Fred and I collected citation information on all the articles published for fifteen years in the top three law reviews. Our central statistical formula had more than fifty variables. Like Epagogix, Fred and I found that seemingly incongruous things mattered a lot. Articles with shorter titles and fewer footnotes were cited significantly more, whereas articles that included an equation or an appendix were cited a lot less. Longer articles were cited more, but the regression formula predicted that citations per page peak for articles that were a whopping fifty-three pages long. (We law professors love to gas on about the law.)

Law review editors who want to maximize their citation rates should also avoid publishing criminal and labor law articles, and focus instead on constitutional law. And they should think about publishing more women. White women were cited 57 percent more often than white men, and minority women were cited more than twice as often. The ultimate merit of an article isn't tied to the race or gender of the author. Yet the regression results suggest that law review editors should think about whether they have been unconsciously setting the acceptance bar unfairly high for women and minority authors whose articles, when published, are cited systematically more often.

Law review editors of course are likely to resist many of these suggestions. Not because they're prima donnas (although believe me, some are), but just because they're human.

A Store of Information

Some long ago when we were taught
That for whatever kind of puzzle you got
You just stick the right formula in
A solution for every fool.

"LEAST COMPLICATED," INDIGO GIRLS

We don't want to be told what to do by a hard-edged and obviously incomplete equation. *Something there is that doesn't love a formula*. Equations, like Robert Frost's walls, limit our freedom to go where we want.

With a little prodding, however, some of our most coveted assessments may yield to the reason of evidence. If this book has succeeded in convincing you that we humans do a poor job in figuring out how much weight to put on various factors when making predictions, then you should be on the lookout for areas in your own job and in your own life where Super Crunching could help.

Stepping back, we can see that technological constraints to data-driven decision making have fallen across the board. The ability to digitalize and store information means that any laptop with access to the Internet can now access libraries several times the size of the library of Alexandria. Computational techniques and fast computers to make the computations were of course necessary, but both regressions and CPUs were in place well before the phenomenon seriously took off. I've suggested here that it is instead our increasing facility with capturing, merging, and storing digital data that has more to do with the current onslaught. It is these breakthroughs in database technology that have also facilitated the commodification of information. Digital data now has market value and it is coalescing into huge data warehouses.

There's no reason to think that the advances in database technology will not continue. Kryder's Law shows no sign of ending. Mashups and merger techniques are becoming automated. Data-scraping programs of the future will not only search the web for new pieces of information but will also automatically seek out the merging equivalents of a Rosetta stone to pair observations from disparate datasets. Increasingly predictive Super Crunching techniques will help mash up the observations from disconnected data.

And maybe most importantly, we should see continued advances in the digital domain's ability to capture information—especially via miniaturized sensors. The miniaturization of electronic sensors has already spurred the capture of all sorts of data. Cell phones are ready to pinpoint owners' whereabouts, purchase soda, or digitally memorialize an image. Never before have so many people had in their pocket an ever-present means to record pictures.

But in the not-too-distant future, nanotechnology may spur an age of "ubiquitous surveillance" in which sensing devices become ever more pervasive in our society. Where retailers now keep track of inventory and sales through collection of data at the checkout scanner, nanotechnology may soon allow them to insert small sensors directly into the product. Nanosensors could keep track of how long you hold on to a particular product before using it, how far you transport it, or whether you will likely use the product in conjunction with other products. Of course, consumers would need to consent to product sensors. But there is no reason to limit the application of nanosensors to embedding them in other objects or clothing. Instead, we may find ourselves surrounded by "smart dust": nanosensors that are free-floating and truly ubiquitous within a particular environment. These sensors might quite literally behave like dust; they would flow through the breeze and, at a size of one cubic millimeter, be nearly undetectable.

The prospect of pervasive digitalization of information is both exciting and scary. It is a cautionary tale for a world without privacy. Indeed, we have seen here several worrisome stories. Poor matching in Florida might have mistakenly purged thousands of African-American voters. Even the story of Epagogix rankles. Isn't art supposed to be determined by the artist? Isn't it better to accept a few cognitive foibles, but to retain more humane environments for creative flourishing? Is Super Crunching good?

CHAPTER 7

Are We Having Fun Yet?

Sandra Kay Daniel, a second-grade teacher at the Emma E. Booker Elementary School in Sarasota, Florida, sits in front of about a dozen second graders. She is a middle-aged matronly African-American woman with a commanding but encouraging voice.

> Open your book up to lesson sixty on page
> 153. And at the count of three. One....
> Two ... Three. Everyone should be on
> page 153. If the yellow paper is going
> to bother you, drop it. Thank you.
> Everyone touch the title of your story.
> Fingers under the title. Get ready to read
> the title.... *The ... Fast ... Way*. We're
> waiting for one member. Thank you.
> Fingers under the title of the story. Get
> ready!
> Class (in unison): "The Pet Goat."

Yes. "The Pet Goat." Fingers under the first word of the story.
　　Get ready to read the story the fast way. GET READY!

The class begins reading the story in unison. As they read, the teacher taps her ruler against the board, beating out a steady rhythm. The students read one word per beat.

Class (to the beat): A girl got a pet goat.
Go on.
Class (to the beat): She liked to go running with her pet goat.
Go on.
Class (to the beat): She played with her . . .
Try it again. Get ready, from the beginning of that sentence. GET
　　READY!
Class (to the beat): She played with her goat in her house.
Go on.
Class (to the beat): The goat ate cans and he ate canes.
Go on.
Class (to the beat): One day her dad said that goat must go.
What's behind the word "said"?
Class (in unison): Comma.
And what does that comma mean?
Class: Slow down.
Let's read that sentence again. Get ready!
Class (to the beat): One day her dad said (pause) that goat
　　must go.
Go on.
Class (to the beat): He eats too many things.
Go on.
Class (to the beat): The girl said that if you let the goat stay with
　　us I will see that he stops eating all those things.
Nice and loud, crisp voices. Let's go.
Class (to the beat): Her dad said that he will try it.
Go on.

Class (to the beat): But one day a car robber came to the girl's
house.

Go on.

Class (to the beat): He saw a big red car in the house and said I
will steal that car.

Go on.

Class (to the beat): He ran to the car and started to open the door.

Go on.

Class (to the beat): The girl and the goat were playing in the
backyard.

Go on.

Class (to the beat): They did not see the car robber. More to come.

More to come? This is a real cliff-hanger. Will the goat stop the car
robber? Will the dad get fed up and kick the goat out?

Millions of us have actually seen Ms. Daniel's class. However, in
the videotape, our attention was centered not on the teacher or the students but on a special guest who was visiting that day. The special
guest, who was sitting quietly by Ms. Daniel's side, was President
George W. Bush.

The videotape of the class was a central scene in Michael Moore's
Fahrenheit 9/11. Just as Ms. Daniel was asking her students to "open
your book to lesson sixty," Andrew Card, the president's chief of staff,
came over and whispered into Bush's ear, "A second plane hit the second tower. America is under attack."

Moore's purpose was to criticize Bush for not paying more attention to what was happening outside the classroom. Yet what was
happening inside Ms. Daniel's classroom concerns one of the fiercest
battles raging about how best to teach schoolchildren. Bush brought
the press to visit this class because Ms. Daniel was using a controversial, but highly effective, teaching method called "Direct Instruction"
(DI).

The fight over whether to use DI, like the fight over evidence-based
medicine, is at heart a struggle about whether to defer to the results of

Super Crunching. Are we willing to follow a treatment that we don't like, but which has been shown in statistical testing to be effective?

Direct Instruction forces teachers to follow a script. The entire lesson—the instructions ("Put your finger under the first word."), the questions ("What does that comma mean?"), and the prompts ("Go on.")—is written out in the teacher's instruction manual. The idea is to force the teacher to present information in easily digestible, bite-size concepts, and to make sure that the information is actually digested.

Each student is called upon to give up to ten responses each minute. How can a single teacher pull this off? The trick is to keep a quick pace and to have the students answer in unison. The script asks the students to "get ready" to give their answers and then after a signal from the teacher, the class responds simultaneously. Every student is literally on call for just about every question.

Direct Instruction also requires fairly small groups of five to ten students of similar skill levels. Small group sizes make it harder for students to fake that they're answering and it lets the teacher from time to time ask individual students to respond, if the teacher is concerned that someone is falling behind.

The high-speed call and response of a DI class is both a challenging and draining experience. As a law professor, it sounds to me like the Socratic method run amok. Most grade schoolers can only handle a couple of hours a day of being constantly on call.

The DI approach is the brainchild of Siegfried "Zig" Engelmann, who started studying how best to teach reading at the University of Illinois in the 1960s. He has written over 1,000 short books in the "Pet Goat" genre. Engelmann, now in his seventies, is a disarming and refreshingly blunt academic who for decades has been waging war against the great minds of education.

Followers of the Swiss developmental psychologist Jean Piaget have championed child-centered approaches to education that tailor the curriculum to the desires and interests of individual students. Followers of the MIT linguist and polymath Noam Chomsky have promoted a whole-language approach to language acquisition. Instead of breaking reading into finite bits of information in order to teach kids specific phonic skills, the whole-language approach embraces a holistic immersion in listening to and eventually reading entire sentences.

Engelmann flatly rejects both the child-centered and whole-language approaches. He isn't nearly as famous as Chomsky or Piaget, but he has a secret weapon—data. Super Crunching doesn't say on a line-by-line basis what should be included in Zig's scripts, but Super Crunching on the back end tells him what approaches actually help students learn. Engelmann rails against educational policies that are the product of top-down philosophizing instead of a bottom-up attention to what works. "Decision makers don't choose a plan because they know it works," he says. "They choose a plan because it's consistent with their vision of what they think kids should do." Most educators, he says, seem to have "a greater investment in romantic notions about children" than they do "in the gritty detail of actual practice or the fact that some things work well."

Engelmann is a thorough pragmatist. He started out in his twenties as an advertising exec who tried to figure out how often you had to repeat an ad to get the message to stick. He kept asking the "Does it work?" question when he turned his attention to education.

The evidence that DI works goes all the way back to 1967. Lyndon Johnson, as part of his War on Poverty, wanted to "follow through" on the vanishing gains seen from Head Start. Concerned that "poor children tend to do poorly in school," the Office of Education and the Office of Economic Opportunity sought to determine what types of education models could best break this cycle of failure. The result was Project Follow Through, an ambitious effort that studied 79,000 children in 180 low-income communities for twenty years at a price tag of more than $600 million. It is a lot easier to Super Crunch when you have this kind of sustained support behind you. At the time, it was the largest education study ever done. Project Follow Through looked at the impact of seventeen different teaching methods, ranging from models like DI, where lesson plans are carefully scripted, to un-structured models where students themselves direct their learning by selecting what and how they will study. Some models, like DI, empha-sized acquisition of basic skills like vocabulary and arithmetic, others emphasized higher-order thinking and problem-solving skills, and still others emphasized positive attitudes toward learning and self-esteem. Project Follow Through's designers wanted to know which model performed the best, not only in developing skills in its area of emphasis, but also across the board.

Direct Instruction won hands down. Education writer Richard Nadler summed it up this way: "When the testing was over, students in DI classrooms had placed first in reading, first in math, first in spelling, and first in language. No other model came close." And DI's dominance wasn't just in basic skill acquisition. DI students could also more easily answer questions that required higher-order thinking. For example, DI students performed better on tests evaluating their abil-ity to determine the meaning of an unfamiliar word from the sur-rounding context. DI students were also able to identify the most appropriate pieces to fill in gaps left in mathematical and visual pat-terns. DI even did better in promoting students' self-esteem than sev-eral child-centered approaches. This is particularly striking because a central purpose of child-centered teaching is to promote self-esteem by engaging children and making them the authors of their own edu-cation.

More recent studies by both the American Federation of Teachers and the American Institutes for Research reviewed data on two dozen "whole school" reforms and found once again that the Direct Instruction model had the strongest empirical support. In 1998, the American Federation of Teachers included DI among six "promising programs to increase student achievement." The study concluded that when DI is properly implemented, the "results are stunning," with DI students outperforming control students along every academic measure. In 2006, the American Institutes for Research rated DI as one of the top two out of more than twenty comprehensive school reform programs. DI again outperformed traditional education programs in both reading and math.

"Traditionalists die over this," Engelmann said. "But in terms of data we whump the daylights out of them."

But wait—it gets even better. Direct Instruction is particularly effective at helping kids who are reading below grade level. Economically disadvantaged students and minorities thrive under DI instruction. And maybe most importantly, DI is scalable. Its success isn't contingent on the personality of some über-teacher. DI classes are entirely scripted. You don't need to be a genius to be an effective DI teacher. DI can be implemented in dozens upon dozens of classrooms with just ordinary teachers. You just need to be able to follow the script.

If you have a school where third graders year after year are reading at a first-grade level, they are seriously at risk of being left behind. DI gives them a realistic shot of getting back to grade. If the school adopts DI from day one of kindergarten, the kids are much less likely to fall behind in the first place.

Imagine that. Engelmann has a validated and imminently replicable program that can help at-risk students. You'd think schools would be beating a path to his door.

What Am I, a Potted Plant?

Direct Instruction has faced severe opposition from educators on the ground. They criticize the script as turning teachers into robots, and for striving to make education "teacher proof."

Can you blame them for resisting? Would *you* want to have to follow a script most of your working day, repeating ad nauseam stale words of encouragement and correction? Most teachers are taught that they should be creative. It is a stock movie genre to show teachers getting through to kids with unusual and idiosyncratic techniques (think

To Sir with Love, Stand and Deliver, Music of the Heart, Mr. Holland's Opus). No one's going to make a motivational drama about Direct Instruction.

Engelmann admits that teacher resistance is a problem. "Teachers initially think this is horrible," he said. "They think it is confining. It's counter to everything they've ever been taught. But within a couple of months, they realize that they are able to teach kids things that they've tried to teach before and never been able to teach."

Direct Instruction caused a minor schism when it was introduced into Arundel Elementary in 1996. Arundel Elementary is perched upon a hill in Baltimore's struggling Cherry Hill neighborhood. It is surrounded by housing projects and apartment complexes. Ninety-five percent of its students are poor enough to qualify for federally subsidized lunches. When Arundel adopted DI, several teachers were so frustrated with the script that they transferred to other schools. The teachers who stayed, though, have come to embrace the system. Matthew Carpenter teaches DI seven hours a day. "I like the structure," he said. "I think it's good for this group of kids."

Most readers of this book probably couldn't abide the idea of having to follow a script hour after hour. Still, there is a joy is seeing your students learn. And a public school teacher confided in me that some of her colleagues liked it for a very mundane reason: "Zero prep," she said. That's right, instead of having to plan your own class lesson day after day, DI instructors can walk into class, open the book, and read, "Good morning, class . . ."

Engelmann's website is clear, if somewhat diplomatic, in emphasizing that teacher discretion is reduced by the Direct Instruction method. "The popular valuing of teacher creativity and autonomy as high priorities must give way to a willingness to follow certain carefully prescribed instructional practices," reads the DI website. Engelmann puts the matter more bluntly: "We don't give a damn what the teacher thinks, what the teacher feels," he said. "On the teachers' own time they can hate it. We don't care, as long as they do it."

The Empire Strikes Back

Engelmann also faces resistance from the academic establishment. The education community is largely unified in their opposition to Direct Instruction. Ignoring the data, they argue that DI doesn't teach high-order thinking, thwarts creativity, and is not consistent with developmental practices.

Opponents argue that DI's strict methodology does not promote learning so much as prompting students to robotically repeat stock answers to scripted questions. They contend that while students learn to memorize responses to questions they expect, students are not prepared to apply this base knowledge to new situations. DI's critics also express concern that its structured approach, with tedious drills and repetition, stifles both student and teacher creativity. They argue that the method treats students as automatons leaving little room for individual thinking. These criticisms, however, ignore the possibility—supported by evidence from standardized tests—that DI equips students who have acquired a stronger set of basic skills with a greater capacity to build and develop creativity. Teachers interviewed after implementation of DI in Broward County, Florida, said the "approach actually allowed more creativity, because a framework was in place within which to innovate," and added that classroom innovation and experimentation were a lot easier once DI had helped students acquire the necessary skills.

Lastly, critics try to discredit DI by arguing that DI causes antisocial behavior. At public meetings, whenever the possibility of switching to DI is mentioned, someone is sure to bring up a Michigan study claiming that students who are taught with DI are more likely to be arrested in their adolescent years. Here's evidence, they say, that DI is dangerous. The problem is that this randomized study was based on the experience of just sixty-eight students. And the students in the DI and the control groups were not similar.

In the end, the Michigan study is just window dressing. The education establishment is wedded to its pet theories regardless of what the evidence says. Education theorist and developer of the *Success for All* teaching model Robert Slavin puts it this way: "Research or no research, many schools would say that's just not a program that fits with their philosophy." For many in the education establishment, philosophy trumps results.

The Bush administration, however, begs to differ. The 2001 "No Child Left Behind" law (NCLB) mandates that only "scientifically based" educational programs are eligible for federal funding. The NCLB statute uses the term "scientifically based research" more than one hundred times. To qualify as "scientifically based," research must "draw on observation or experiment" and "involve rigorous data analyses that are adequate to test the stated hypotheses." This is the kind of stuff that would make any Super Cruncher salivate. Finally a fair fight, where the education model that teaches the best will prevail.

Bush's education advisors have been taking the mandate quite seriously. The Department of Education has taken the lead, spending more than $5 billion in funding randomized testing and funding evidence-based literature reviews to assess the state of knowledge of "what works." As *Fahrenheit 9/11* shows, Bush is personally flogging the effectiveness of Direct Instruction.

On the ground, however, the requirement that states adopt scientifically based methods has not worked a sea change on the education environment. State education boards currently tend to require textbooks and materials to be a mishmash of elements that individually are supposed to be scientifically based. A "balanced literacy" approach, which mixes elements of phonetic awareness as well as holistic experiences in reading and writing, is now in the ascendancy. California requires that primary reading materials contain a mixture of broad features.

Ironically, NCLB's requirement of "scientifically based" methods has become the catalyst for *excluding* Direct Instruction from many states' approved lists because it does not contain some holistic elements. There are no good studies indicating that "balanced learning" materials perform as well as Direct Instruction, but that doesn't keep states from disqualifying DI as even an option for local school adoption. At the moment, Direct Instruction, the oldest and most validated program, has captured only a little more than 1 percent of the grade-school market. Will this share rise as the empirical commands of NCLB are more fully realized? In the immortal words of "The Pet Goat," "more to come."

The Status Squeeze

The story of Engelmann's struggle with the educational establishment raises once again the core themes of this book. We see the struggle of intuition, personal experience, and philosophical inclination waging war against the brute force of numbers. Engelmann for decades has staked out the leading edge of the Super Cruncher's camp. "Intuition is perhaps your worst enemy," Engelmann said, "if you want to be smart in the instructional arena. You have to look at the kid's performance."

In part, the struggle in education is a struggle over power. The education establishment and the teacher on the line want to keep their authority to decide what happens in the classroom. Engelmann and the mandate of "scientifically based" research are a direct threat to that power. Teachers in the classroom realize that their freedom and discretion to innovate is threatened. Under Direct Instruction, it is Zig who runs the show, who sets up the algorithm, who tests which script works best.

It's not just the teacher's power and discretion that is at stake. Status and power often go hand in hand. The rise of Super Crunching threatens the status and respectability of many traditional jobs.

Take the lowly loan officer. Once, being a loan officer for a bank was a moderately high status position. Loan officers were well paid and had real power to decide who did and did not qualify for loans. They were disproportionately white and male.

Today, loan decisions are instead made at a central office based on the results of a statistical algorithm. Banks started learning that giving loan officers discretion was bad business. It's not just that officers used this discretion to help their friends, or to unconsciously (or consciously) discriminate against minorities. It turns out that looking a customer in the eye and establishing a relationship doesn't help predict whether or not the customer will really repay the loan.

Bank loan officers, stripped of their discretion, have become nothing more than glorified secretaries. They literally just type in applicant data and click send. It's little wonder that their status and salaries have plummeted (and officers are much less likely to be white men). In education, the struggle between the intuitivists and the Super Crunchers is ongoing, but in consumer lending the battle ended long ago.

Following some other guy's script or algorithm may not make for the most interesting job, but time and time again it leads to a more effective business model. We are living in an age where dispersed discretion is on the wane. This is not the end of discretion; it's the shift of discretion from line employees to the much more centralized staff of Super Crunching higher-ups. Line employees increasingly feel like "potted plant" functionaries who are literally told to follow a script. Marx was wrong about a lot of things, but through a Super Crunching lens, he looks downright prescient when he said that the development of capitalism would increasingly alienate workers from their work-product.

These algorithm-driven scripts have even played a role in the outsourcing movement. Once discretion is stripped from line employees, they don't need to be as skilled. A pretested script is a cheaper way to lead customers through a service problem or to upsell related products—and it's even cheaper if the script is read by someone sitting in a Third-World call center. Some individual salespeople using their intuition and experience may in fact be legitimately outstanding, but if you're running a large-scale operation selling relatively homogenous products, you're going to do a heck of a lot better if you can just get your staff to stick with a tried-and-true script.

The shift of discretion and status from traditional experts to database oracles is also happening in medicine. Physicians report that patients now often treat them merely as alternative sources of information. Patients demand, "Show me the study." They want to see the study that says chemo is better than radiation for stage-three lung cancer. Savvy patients are treating their doctors less like 1970s television icon Marcus Welby, and more like a human substitute for a web portal. The physician is merely the conduit of information.

The rise of evidence-based medicine is changing our very conception of what doctors are. "It is a signal that in medicine," Canadian internist Kevin Patterson laments, "ours is a less heroic age."

"So the warriors are being replaced by the accountants," Patterson said. "Accountants know the whole world thinks their lives are gray—demeaned by all that addition. Doctors aren't used to thinking of themselves that way. But in the real world, where numbers matter, accountants know how powerful they are."

Physician status is in decline. People are looking past the M.D.s, who merely disseminate information, and toward the Ph.D.s, who create the database to discover information. While a graduate student in sciences has to actually create information in his or her thesis to get a Ph.D., med students only have to memorize other people's information (including how to do certain procedures). In a world where information is sovereign, there may come a time when we ask, "Are you a real doctor, or just a physician?"

Or maybe not. Respect doesn't necessarily come with power. Society is used to revering sage intuitivists. It can bow down to the theoretical genius of an Einstein or a Salk, but it is harder to revere the number-crunching "accountants" who tell us the probability that our cancer will respond to chemotherapy is 37.8 percent. In the movie *Along Came Polly,* Ben Stiller plays your typical actuarial gearhead. He's the kind of guy who's afraid to eat bar peanuts because "on average only one out of every six people wash their hands when they go to the bathroom." His character leads a small, circumscribed life that is devoid of passion. He may wield power, but he doesn't claim our respect. Power and discretion are definitely shifting from the periphery to the Super Crunching center. But that doesn't mean Super Crunchers are going to find they have an easier time dating.

Would You Buy a Used Car from a Super Cruncher?

Even in areas where number crunching improves the quality of advice, it can sometimes perversely undermine the public's confidence. The heroic conception of expertise was that of an expert giving settled answers. People are more likely to think of statistics as infinitely malleable and subject to manipulation. (Think of the "Lies, damn lies and statistics!" warning.)

This is a more precise, but less certain world. The classical conception of probability is a world of absolutes. To the classicist, the probability of my currently having prostate cancer is either 0 or 100 percent. But we are all frequentists now. Experts used to say "yes" or "no." Now we have to contend with estimates and probabilities.

Super Crunching thus affects us not just as employees but also as consumers and clients as well. We are the patients who demand to see the study. We are the students who are forced to learn *The...Fast... Way*.™ We are the customers who are upsold by a statistically validated (outsourced) script.

Many of the Super Crunching stories are examples of unmitigated consumer progress. Offermatica helps improve your surfing experience by figuring out what websites work the best. Thanks to Super Crunching we now know that targeted job search assistance is a lot more effective than financial incentives in getting unemployed workers back on the job. Physicians may dislike the reduction in their status and power, but at the end of the day medicine should be about saving lives. And for many serious medical risks, it is the database analysis of scientists that points toward progress.

Super Crunching approaches are winning the day and driving out intuition and experience-based expertise because Super Crunching improves firms' bottom-line profitability—usually by enhancing the consumer's experience. A seller that can predict what you want to buy can make life easier. Whether it's Amazon's "Customers who bought this, also bought this" feature or Capital One's validated upselling, or Google's heavily crunched Gmail ads, the bottom line is an improvement in quality. Statistical software is even in place to tell you what not to buy. Peapod, the online grocery store, will interrupt my online session to ask, "Do you really want to buy twelve lemons?" because they know that's unusual and they'd prefer to catch mistakes early and keep me a happy customer.

Epagogix Agonistes

Notwithstanding these benefits, there persists a lingering concern that the Super Crunching of product attributes will lead to a grinding uniformity. The scripted performances of Direct Instruction teachers and CapOne sales reps are not just wearing on the employees; they can wear on the audience as well. Epagogix's meddling with movie scripts is even more troubling.

The story of Epagogix's interference with the movie business can be seen as the death of art. A major Hollywood figure recently told the head of Epagogix, "You know, you absolutely revolt me." Copaken told me, "He kept calling me Jim Jones and saying that I wanted him to drink the Kool-Aid." We have begun to live in a world where a statistical formula tells authors what to put in their scripts. Sorry, Mr. Ivory, the hero needs a buddy if you want us to make your movie.

This truly is a scary picture of the future—where the artist is handcuffed by the gearhead. Yet this concern ignores the other shackles that are already in place. The commercialization train left the station long, long ago. Studios have been tampering with artistic vision for decades in an effort to increase movie sales.

The biggest problem isn't that studios have been interfering; rather, it's that they've been doing it badly. I'm more scared of studio execs wielding their greenlight power based on nothing more than intuition and experience than I am of a formula. Epagogix doesn't represent a shift of power from the artist to the gearhead as much as it does a shift of power from the overconfident studio apparatchiks to people who can make dramatically more reliable interventions. When Copaken recently suggested to a powerful producer that neural predictions might be more objective because they don't have to worry about bruising the egos of stars, the producer responded, "I can be just as objective as any computer." Copaken suggested that the producer "may be subconsciously affected by his responsibility and opportunities within the industry in a way that Epagogix is not."

There will always be legitimate and ultimately irresolvable tensions between artistic and commercial goals. However, there should be no disagreement that it's a tragedy to *mistakenly* interfere. If a studio is going to change a writer's vision in the name of profitability, it should be confident that it's right. Epagogix is moving us toward evidence-based interference.

It's also a move toward meritocracy. The Hollywood star writer system gives inordinate weight to writers who've had a hit movie in the past. It's very hard for a newcomer to even get read. Epagogix democratizes the competition. If you have a script that's off the predictive charts, you're going to have a lot better chance of seeing it on the screen whether you're a "proven" writer or not. Even some successful writers have embraced the Epagogix method. Dick Copaken told me about a well-known writer who sought out Epagogix's help because he's trying to make the shift to directing. "He figures the best way to get a job directing," Copaken said, "is to write a mega-successful script."

But oh, the humanity. Imagine the crushing uniformity, the critics remonstrate, that would be produced by this brave new world of film by formula. Again, this concern ignores the present pressures toward commercial conformity. Epagogix's formula didn't create the idea of a formulaic movie. It is entirely possible that an Epagogix world of cinema would exhibit more diversity than the present marketplace.

Epagogix does not use a simple cookie-cutter formula. Its neural network takes into account literally hundreds of variables, and their impact on the revenue predictions are massively interdependent. Moreover, the neural network is constantly retraining itself. Of course, so are the experiential experts at the studios. But this is precisely the horse race of figuring out the correct weights that humans are bound to lose. The studio expert is much more likely to fall back on simpler rules of thumb that lead to even less variety.

Epagogix's financial success can actually facilitate more experimentation. If it's really true that the neural network can help raise a studio's batting average from .300 to .600, studios might have more flexibility to pursue riskier or unusual projects. All the extra cash from improved predictability might give studios more wiggle room to experiment. And while Epagogix is a great leap forward over the experientialist mode of prediction, its predictions are still bound by history. The Super Cruncher studios of tomorrow will also manufacture their own new data by experimenting.

The Super Crunching of art seems perverse, but it also represents an empowerment of the consumer. Epagogix's neural network is helping studios predict what qualities of movies consumers will actually like. It thus represents a shift of power from the artist/seller to the audience/consumer. Epagogix, from this perspective, is part and parcel of the larger tendency of Super Crunching to enhance consumer quality. Quality, like beauty, is in the eye of the beholder, and Super Crunching helps to match consumers with products and services that they'll find beautiful.

Beware of Super Crunchers Bearing Gifts

Everybody loves a freebie—you know, those little gifts that sellers send their best clients. Still, we should be worried if we see a seller treating us better than its other customers. In a world of Super Crunching, sellers' promotions are far from random. When Amazon sends you a nice desk ornament out of the blue, your first reaction should now be "Yikes, I've been paying too much for my books."

When firms Super Crunch on quality, they tend to help consumers. However, when firms Super Crunch on price, hold on to your wallet. The dark side of customer relations management is firms trying to figure out just how much money they can squeeze out of you and still keep your business. In the old days, the firm's lack of pricing sophistication protected us from a lot of these shenanigans.

Nowadays more and more firms are going to be predicting their customers' "pain points." They are becoming more adept at figuring out how much pricing pain individual consumers are willing to endure and still come back for more. More and more grocery stores are calculating their customers' pain points. It would be a scandal if we learned that your local Piggly Wiggly was charging customers different prices for the same jar of peanut butter. However, there is nothing to stop them from setting individualized coupon amounts that they think are the minimum discount to get you to buy. At the checkout aisle, after they have just scanned in all that information about you (including swiping your loyalty card), they can print out tailored coupons with prices just for you. This new predictive art is a weird twist on the Clinton line "I feel your pain." They feel your pain all right; but they experience it as pleasure because the high net price to you is pure profit to them.

In a world of Super Crunching, it's going to be a lot harder to rely on other consumers to keep your price in line. The fact that price-conscious buyers patronize a store is no longer an indication that it will be a good place for you, too. The nimble number cruncher will be able to size you up in a few nanoseconds and say, "For you, the price is . . ." This is a new kind of *caveat emptor,* where consumers are going to have to search more to make sure that the offered price is fair. Consumers are going to have to engage in a kind of number crunching of their own, creating and comparing datasets of (quality-adjusted) competitive prices. This is a daunting prospect for people like me who are commercially lethargic by nature. Yet the same digitalization revolution that has catalyzed seller crunching has also been a boon to buy-side analysis. Firms like Farecast.com, E-loan, Priceline, and Realrate.com allow customers to comparison shop more easily. In effect, they do the heavy lifting for you and help level the playing field with the price-crunching sellers. For consumers worried about the impact of Super Crunching on price, it is both the best of times and the worst of times.

Discrimination, by Other Means

The prospect of increased price discrimination is scary enough. Even more disturbing is the notion that Super Crunching can also be used to facilitate racial discrimination. Earlier I spoke about the uncontroversial successes of data-driven lending decisions. It really is true that statistical formulas beat the pants off any discretionary system of loan officers. In part, this is because statistical formulas don't have feelings. Regression equations, unlike flesh-and-blood loan officers, cannot harbor racial animus. So the seismic shift toward centralized statistical lending and insurance decisions has largely disabled hatred as a motive for minority loan or insurance denials.

However, the shift to statistical decision making has not been a civil rights panacea. The algorithmic lending and insuring policies open the possibility for race to influence centralized policies. It is highly unlikely that the algorithm would be expressly contingent on race. It's simply too likely that a race-contingent formula would become publicly known. Yet the algorithms that are formally race-neutral have at times been challenged for facilitating a type of virtual redlining. Geographic redlining was the historic practice of refusing to lend in minority neighborhoods. Virtual redlining is the analogous practice of refusing to lend to any database group that has too many minorities. The worry here is that lenders can mine a database to find characteristics that strongly correlate with race and use those characteristics as a pretext for loan denials. Members of minority groups have challenged lending denials that rely on factors like the small size of a loan or a borrower's poor credit history as pretextual characteristics that highly correlate with race.

Our civil rights statutes prohibit race-contingent lending policies. Even if a lender found that Hispanics were more likely to default than Anglos with the same credit score, the lender could not legally condition its lending decisions on race. The lender may be tempted, however, to use Super Crunching to end-run the civil rights prohibition. As long as it uses a race-neutral means, it will be very hard to establish that race was an underlying, illicit motivation for the policy.

Such virtual redlining may also take place in the insurance context. An African-American woman, Chikeitha Owens, who was denied homeowner's insurance coverage due to her poor credit, sued Nationwide Insurance. She claimed that the company's use of her credit score history effectively created a racialized category which denied coverage to applicants who were otherwise qualified.

In fact, when it comes to affirmative action, the Supreme Court invites this kind of end run. Justice Sandra Day O'Connor as the swing vote for the Court in a series of crucial affirmative action opinions said that decision makers had to try to find "race-neutral means to increase minority participation" before implementing an affirmative action policy that was expressly contingent on race. Some schools have responded to the invitation by searching for race-neutral criteria for admission that disproportionately favor minority applicants. Some California schools, for example, now favor applicants whose mothers have not graduated from college. This admissions criterion is explicitly motivated by a goal of increasing minority enrollment. Yet Justice O'Connor's standard is an open invitation to more elaborate formulas for predicting race. In a world where express race preference is prohibited, Super Crunching opens the possibility for conditioning behavior on predicted race.

Probabilistically Public

The idea that a university or insurer could predict your race is itself just another way that Super Crunching is reducing our sphere of effective privacy. Suddenly we live in a world where less and less is hidden about who we are, what we have done, and what we will do.

Part of the privacy problem isn't a problem of Super Crunching; it's the dark side of digitalization. Information is not only easier to capture now in digital form, but it is also virtually costless to copy. It's scary to live in a world where ChoicePoint and other data aggregators know so much about us. There is the legitimate fear that information will leak. In May 2006, this fear became real for more than 17.5 million military veterans when electronic records containing their Social Security numbers and birth dates were stolen from a government employee's home. The government told veterans to be "extra vigilant" when monitoring bank and credit card records, but the risk of identity theft remains. And this risk is not just limited to bureaucratic mishaps. A laptop was stolen from the home of a Fidelity Investments employee and—poof, the personal information of 196,000 HP employees was suddenly up for grabs. Or an underling at AOL hit the wrong key and poof, the personal search information of millions of users was released onto the net.

We're used to giving the phone company or websites passwords or answers to challenge questions so that they can verify that we are who we say we are when we call in. But today, new services like Corillian are providing retailers with challenge questions and answers that you have never provided, in a matter of seconds. You might walk into Macy's to apply for a credit card and while you're waiting, Macy's will ask you where your mom lived in 1972. Statistical matching algorithms let Super Crunchers connect widely disparate types of data and ferret out facts in nanoseconds that would have taken weeks of effort in the past.

Our privacy laws so far have been mainly concerned with protecting people's privacy in their homes and curtilage (the enclosed area in land surrounding your house). To Robert Frost, home was the place where, "when you have to go there, they have to take you in." To the Constitution, however, home is quintessentially the place where you have a "reasonable expectation of privacy." Out on the street, the law says we don't expect our actions to be private and the police are free without a warrant, say, to listen in on our conversations.

The law in the past didn't need to worry much about our walking-around privacy, because out on the street we were usually effectively anonymous. Donald Trump may have difficulty going for a walk incognito in New York, but most of us could happily walk across the length and breadth of Manhattan without being recognized.

Yet the sphere of public anonymity is shrinking. With just a name, we can google people on the fly to pull up addresses, photographs, and myriad pieces of other information. And with face-recognition software, we don't even need a name. It will soon be possible to passively identify passersby. The first face-recognition applications were operated by police—looking for people with outstanding warrants at the Super Bowl. In Massachusetts, police recently were able to use face-recognition software to catch Robert Howell, a fugitive featured on the television show *America's Most Wanted*. Law enforcement officials tracked him down after they used his mug shot from the TV show to find a match in a database of over nine million digital driver's license photos. Although Howell had managed to obtain a driver's license under a different name, facial recognition software eventually caught up with him. This Super Crunching software is also being used to catch people who fraudulently apply for multiple licenses under different names. While not perfect, the database predictions were accurate enough to flag more than 150 twins as potential fraud cases.

In the Steven Spielberg movie *Minority Report,* Tom Cruise's character was bombarded with personalized electronic ads that recognized and called out to him as he walked through a mall. For the moment, this is still the stuff of science fiction. But we're getting closer. Now passive identification is coming to the web. PolarRose.com is using face recognition to improve the quality of image searches. Google image searches currently rely on the text that appears near web images. PolarRose, on the other hand, creates 3-D renditions of the faces, codes for ninety different facial attributes, and then matches the image to an ever-growing database. Suddenly, if you happened to be walking by in the background when a tourist snaps a picture, the whole world could learn where you were. Any photo that is subsequently posted to websites like flickr.com could reveal your whereabouts.

Most of the popular discussion of face recognition emphasizes the software that needs to successfully "code" different facial attributes. But make no mistake, facial recognition is Super Crunching looking for high-probability predictions. And once you've been identified, the Super Crunching cat is out of the bag, as increasing numbers of people will be able to determine what library books you forgot to return, which politicians you gave money to, what real estate you own, and countless other bytes of data about you. Walking into H&H and buying a bagel is technically public, but for most of us non-celebrities, public anonymity allowed for a vast range of unobserved freedom of movement. Super Crunching is reducing the sphere of being private in public.

Sherlock Holmes was famous for deducing intricate details about a person's past by observing just a few details of their present. But given access to much more intricate datasets, Super Crunchers can put Holmesian prediction to shame. Given these 250 variables, there's a 93 percent chance that you voted for Nader. It's elementary, my dear Watson.

Data mining does more than give new meaning to "I know what you did last summer." Super Crunchers can also make astute predictions about what you will do *next* summer. Traditionally, the right to privacy has been about preserving past and present information. There was no need to worry about keeping future information private. Since future information doesn't exist yet, there was nothing to keep private. Yet data-mining predictions raise just this concern. Super Crunching in a sense puts our future privacy at risk because it can probabilistically predict what we will do. Super Crunching moves us toward a kind of statistical predeterminism.

The 1997 sci-fi thriller *Gattaca* imagined a world in which genetics was destiny. The hero's parents were told at his birth that he had a 42 percent chance of manic depression and life expectancy of 34.2 years. But right now it's possible for Super Crunchers to look at a collection of innocuous past behaviors and make chillingly accurate assessments of the future. For example, it's a little unnerving to think that Visa, with a little mining of my credit card charges, can make a pretty accurate guess of whether I'll divorce in the next five years.

"Data huggers"—the people who are scared about the untoward uses of public data—have a lot to be worried about. Google's corporate mission is "to organize the world's information and make it universally accessible and useful." This ambitious goal is seductively attractive. However, it is not counterbalanced by any concern for privacy. Data-driven predictions are creating new dimensions where our past and even future actions will be "universally accessible."

The slow erosion of the private sphere makes it harder to realize what is happening and rally resistance. Like a frog slowly boiling to death, we don't notice that our environment is changing. People in Israel now expect to be repeatedly checked by a metal detector as they go about their daily tasks. Sometimes incremental steps of "progress" take us down a path where collectively we end up eating hot-house tomatoes and Wonder Bread that are at best the semblance of food. The fear is that number crunching will somehow similarly degrade our lives.

Newspaper reporters feel compelled to quote privacy pundits who raise concerns and call for debate. Yet most people, when it comes down to it, don't seem to value their privacy very much. The Ponemon Institute estimates that only 7 percent of Americans change their behaviors to preserve their privacy. Rare is the person who will reject the EZ-pass system (and its discount) because it can track car movements. Carnegie-Mellon economist Alessandro Acquisti has found that people are happy to surrender their Social Security number for just a fifty-cents-off coupon. Individually, we're willing to sell our privacy. Sun's founder and CEO Scott McNealy famously declared in 1999 that we "have no privacy—get over it." Many of us already have.

Super Crunching affects us not only as customers and as employees but also as citizens. I, for one, am not worried about people googling me or predicting my actions. The benefits of indexing and crunching the world's information far outweigh its costs. Other citizens may reasonably disagree. One thing is for certain: consumer pressure by itself is not likely to restrain the Super Crunching onslaught. The data huggers of the world need to unite (and convince Congress) if the excesses of data-based decision making are going to be constrained.

Truth is often a defense. But even true predictions at times may hurt customers and employees if they allow firms to take advantage of us, and predictions can hurt us as citizens if they allow others to inappropriately invade our past, present, or future privacy. The larger concern is about inaccurate (untrue) predictions. Without appropriate protections, they can hurt everybody.

Who Is John Lott?

On September 23, 2002, Mary Rosh posted to the web a rather harsh criticism of an empirical paper that I wrote with my colleague John Donohue. Rosh said:

> The Ayres and Donohue piece is a joke. I saw it a while ago.... A friend at the Harvard Law School said that Donohue gave the paper there and he was demolished...

The article that Rosh was criticizing was about the impact of concealed handgun laws on crime. It was a response to John Lott's "More Guns, Less Crime" claim. Lott created a huge dataset to analyze the impact that concealed weapon laws had on crime. His startling finding was that states which passed laws making it easy for law-abiding citizens to carry concealed weapons experienced a substantial decrease in crime. Lott believed that criminals would be less likely to commit crime if they couldn't be sure whether or not their victims were armed.

Donohue and I took Lott's data and ran thousands of regressions exploring the same issue. Our article refuted Lott's central claim. In fact, we found twice as many states that experienced a statistically significant *increase* in crime after passage of the law. Overall, however, we found that the changes were not substantial, and these concealed weapon laws might not impact crime one way or the other.

That's when Mary Rosh weighed in on the web. Her comment isn't so remarkable for its content—that's part of the rough and tumble of academic disputes. The comment is remarkable because Mary Rosh is really John Lott. Mary Rosh was a "sock puppet" pseudonym (based on the first two letters of his four sons' names). Lott as Rosh posted dozens upon dozens of comments to the web praising his own merits and slamming the work of his opponents. Rosh, for example, identified herself as a former student of Lott's and extolled Lott's teaching. "I have to say that he was the best professor that I ever had," she wrote. "You wouldn't know that he was a 'right-wing ideologue' from the class."

Lott is a complicated and tortured soul. He is often the smartest guy in the room. He comes to seminars and debates consummately prepared. I first met him at the University of Chicago when I was delivering a statistical paper on New Haven bail bondsmen. Lott had not only read my paper carefully, he'd looked up the phone numbers of New Haven bond dealers and called them on the phone. I was blown away.

He is incredibly combative in public, but just as soft-spoken, even meek, when speaking one-on-one. Lott is also a physical presence. He is tall and has striking features—Ichabod Crane–like in their lack of proportion. Mary Rosh has even described him:

> I had Lott as a teacher about a decade ago, and he has a quite noti-
> cable [sic] scar across his forehead. It looked like it cut right

through his eyebrows going the entire width of his forehead. [T]he scar was so extremely noticable [sic] that people talked and joked about it. Some students claimed that he had major surgery when he was a child.

Before the Mary Rosh dissembling, I was instrumental in bringing John to Yale Law School for two years as a research fellow. Make no mistake, John Lott has some serious number-crunching skills.

His concealed-weapon empiricism was quickly picked up by gun-rights advocates and politicians as a reason to oppose efforts at gun control and advance the cause of greater freedom to carry guns. In the same year that Lott's initial article was published, Senator Larry Craig (R-Idaho) introduced The Personal Safety and Community Protection Act, which was designed to facilitate the carrying of concealed firearms by nonresidents of a state who had obtained valid permits to carry such weapons in their home state. Senator Craig argued that the work of John Lott showed that arming the citizenry via laws allowing the carrying of concealed handguns would have a protective effect for the community at large because criminals would find themselves in the line of fire.

Lott has repeatedly been asked to testify to state legislatures in support of concealed gun laws. Since Lott's original article was published in 1998, nine additional states have passed his favored statute. This book is about the impact that Super Crunching is having on real-world decisions. It's hard to know for sure whether Lott's regressions were a but-for cause of these new statutes. Still Lott and his "More Guns/Less Crime" regressions have had the kind of influence that most academics can only dream of.

Lott generously made his dataset available not only to Donohue and me, but to anyone who asked. And we dug in, double-checking the calculations and testing whether his results held up if we slightly changed some of his assumptions. Econometricians call this testing to see whether results are "robust."

We had two big surprises. First, we found that if you made innocuous changes in Lott's regression equation, the crime-reducing impacts of Lott's laws often vanished. More disturbingly, we found that Lott had made a computer mistake in creating some of his underlying data. For example, in many of his regressions, Lott tried to control for whether the crime took place in a particular region (say, the Northeast) in a particular year (say, 1988). But when we looked at his data, many of these variables were mistakenly set to zero. When we estimated his formula on the corrected data, we again found that these laws were more likely to increase the rate of crime.

Let me stress that both of these mistakes are the kind of errors that, but for the grace of God, I or just about any other Super Cruncher might make—especially regarding the coding error. There are literally hundreds of data manipulations that need to be made in getting a large dataset in shape to run a regression. If the gearhead makes a mistake on any one of the transformations, the bottom-line predictions may be inaccurate. I have no concern that Lott purposefully miscoded his data to produce predictions that supported his thesis. Nonetheless, it is disturbing that after Donohue and I pointed out the coding errors, Lott and his coauthors continued to rely on the flawed data. As Donohue and I said in a response to our initial article, "repeatedly bringing erroneous data into the public debate starts suggesting a pattern of behavior that is unlikely to engender support for the Lott ['More Guns/Less Crime'] hypothesis."

We are not the only ones to engage the topic. More than a dozen different authors have exploited the Lott data to reanalyze the issue. In 2004, the National Academy of Science entered into the debate, conducting a review of the empirical research of firearms and violent crime, including Lott's work. Their panel of experts found: "There is no credible evidence that 'right-to-carry' laws, which allow qualified adults to carry concealed handguns, either decrease or increase violent crime." At least for the moment, this pretty much sums up what many academics feel about the issue.

Lott, however, fights on undaunted. Indeed, John is such a tenacious adversary that I'm a little scared to mention his name here in this book. In 2006, Lott took the extraordinary step of suing Steve Levitt for defamation, growing out of a single paragraph in Levitt's bestselling *Freakonomics* book, which said in part:

> Lott's admittedly intriguing hypothesis doesn't seem to be true. When other scholars have tried to replicate his results, they found that right-to-carry laws simply don't bring down crime.

Levitt's endnote supported this claim by citing . . . you guessed it, my article with Donohue that Mary Rosh thought was a joke. Lott's defamation charge all depends on the meaning of "replicate." Lott claims that Levitt was suggesting that Lott falsified his results—that he committed the cardinal sin of "editing the output file." I find it shocking that Lott brought this suit, especially since Donohue and I couldn't replicate some of his results once we corrected Lott's clear coding error (coding errors, by the way, that Lott himself has conceded).

Thankfully, the district court has dismissed the *Freakonomics* claim. Early in 2007, Judge Ruben Castillo found that the term "replicate" was susceptible to non-defaming meanings. The judge pointed to the same Ayres and Donohue endnote, saying that it clarified "the intended definition of the term 'replicate' to be simply that other scholars have disproved Lott's gun theory, not that they proved Lott falsified his data."

But What If It's Wrong?

The Lott saga has important lessons for Super Crunchers. First, Lott should be applauded for his exemplary sharing of data. Even though Lott's reputation has been severely damaged by the Mary Rosh incident and a host of other concerns, Lott's open-access policy has contributed to a new sharing ethic among data crunchers. I, for one, now share data whenever I legally can. And several journals including my own *Journal of Law, Economics, and Organization* now require data sharing (or an explanation why you can't share). Donohue and I would never have been able to evaluate Lott's work if he had not led the way by giving us the dataset that he worked on.

The Lott saga also underscores why it is so important to have independent verification of results. It's so easy for people of good faith to make mistakes. Moreover, once a researcher has spent endless hours producing an interesting result, he or she becomes invested in defending it. I include myself in this tendency. It's easy and true to charge that the intuitivist and experientialist are subject to cognitive biases. Yet the Lott saga shows that empiricists are, too. Lott's adamant defense of his thesis in the face of such overwhelming evidence underscores this fact. Numbers don't have emotions or preferences, but the number crunchers that interpret them do.

My contretemps with Lott suggests the usefulness of setting up a formalized system of empirical devil's advocacy akin to the role of an *Advocatus Diaboli* in the Roman Catholic Church. For over 500 years, the canonization process followed a formal procedure in which one person (a *postulator*) presents the case in favor and another (the *promoter of the faith*) presents the case against. According to Prospero Lamertini (later Pope Benedict XIV [1740–58]):

> It is [the promoter of the faith's duty] to critically examine the life of, and the miracles attributed to, the individual up for sainthood or blessedness. Because his presentation of facts must include everything unfavorable to the candidate, the promoter of the faith is popularly known as the *devil's advocate*. His duty requires him to prepare in writing all possible arguments, even at times seemingly slight, against the raising of any one to the honours of the altar.

Corporate boards could create devil's advocate positions whose job it is to poke holes in pet projects. These professional "No" men could be an antidote to overconfidence bias—without risking their jobs. The Lott story shows that institutionalized counterpunching may also be appropriate for Super Crunchers to make sure that their predictions are robust.

Among academic crunchers, this devil's advocacy is a two-way street. Donohue and I have crunched numbers testing the robustness of Lott's "More Guns" thesis. Lott again and again has recrunched numbers that my coauthors and I have run. Lott challenged the robustness of an article that Levitt and I wrote showing that the hidden transmitter LoJack has a big impact on reducing crime. And Lott has also recrunched numbers to challenge a Donohue and Levitt article showing that the legalization of abortion reduced crime. To my mind, none of Lott's counterpunching crunches has been persuasive. Nonetheless, the real point is that it's not for us or Lott to decide. By opening up the number crunching to contestation, we're more likely to get it right. We keep each other honest.

Contestation and counterpunching is especially important for Super Crunching, because the method leads to centralized decision making. When you are putting all your eggs in a single decisional basket, it's important to try to make sure that the decision is accurate. The carpenter's creed to "measure twice, cut once" applies. Outside of the academy, however, the useful Lott/Ayres/Donohue/Levitt contestation is often lacking. We're used to governmental or corporate committee reports giving the supposedly definitive results of some empirical study. Yet agencies and committees usually don't have empirical checks and balances. Particularly when the underlying data is proprietary or confidential—and this is still often the case with regard to both business and governmental data—it becomes impossible for outsiders like Lott or me to counterpunch. It thus becomes all the more important that these closed Super Crunchers make themselves answerable to loyal opposition within their own organizations. Indeed, I predict that data-quality firms will appear to provide confidential second opinions—just like the big four accounting firms will audit your books. Decision makers shouldn't rely on the word of just one number cruncher.

Most of this book has been chock full of examples where Super Crunchers get it right. We might or might not always like the impact of their predictions on us as consumers, employees, or citizens, but the predictions have tended to be more accurate than those of humans unaided by the power of data mining. Still, the Lott saga underscores the fact that number crunchers are not infallible oracles. We, of course, can and do get it wrong. The world suffers when it relies on bad numbers.

The onslaught of data-based decision making if not monitored (internally or externally) may unleash a wave of mistaken statistical analysis. Some databases do not easily yield up definitive answers. In the policy arena, there are still lively debates about whether (a) the death penalty, or (b) concealed handguns, or (c) abortions reduce crime. Some researchers have so comprehensively tortured the data that their datasets become like prisoners who will tell you anything you want to know. Statistical analysis casts a patina of scientific integrity over a study that can obscure the misuse of mistaken assumptions.

Even randomized studies, the gold-standard of causal testing, may yield distorted predictions. For example, Nobel Prize–winning econometrician James Heckman has appropriately railed against reliance on randomized results where there is a substantial decline in the number of subjects who complete the experiment. For example, at the moment I'm trying to set up a randomized study to test whether Weight Watchers plus a financial incentive to lose weight does better than Weight Watchers alone. The natural way to set this up is to find a bunch of people who are about to start Weight Watchers and then to flip a coin and give half a financial incentive and make the other half the control group. The problem comes when we try to collect the results. The Constitutional prohibition against slavery is a very good thing, but it means that we can't mandate that people will continue participating in our study. There is almost always attrition as some people after a while quit responding to your phone calls. Even though the treatment and control group were probabilistically identical in the beginning, they may be very different by the end. Indeed, in this example, I worry that people who fail to lose weight are more likely to quit the financial incentive group thus leaving me at the end with a self-censored sample of people who have succeeded. That's not a very good test of whether the financial incentive causes more weight loss.

One of the most controversial recent randomized studies concerned an even more basic question: are low-fat diets good for your health? In 2006, the Women's Health Initiative (WHI) reported the results of a $415 million federal study. The researchers randomly assigned nearly 49,000 women ages fifty to seventy-nine to follow a low-fat diet or not, and then followed their health for eight years.

The low-fat diet group "received an intensive behavioral modification program that consisted of eighteen group sessions in the first year and quarterly maintenance sessions thereafter." These women did report eating 10.7 percent less fat at the end of the first year and 8.1 percent less fat at the end of year six. (They also reported eating on average each day an extra serving of vegetables or fruit.)

The shocking news was that, contrary to prior accepted wisdom, the low-fat diet did not improve the women's health. The women assigned to the low-fat diet weighed about the same and had the same rates of breast cancer, colon cancer, and heart disease as those whose diets were unchanged. (There was a slightly lower risk of breast cancer—42 per 10,000 per year in the low-fat diet group, compared with 45 per 10,000 in the regular diet group—but the difference was statistically insignificant.)

Some researchers trumpeted the results as a triumph for evidence-based medicine. For them, this massive randomized trial conclusively refuted earlier studies which suggested that low-fat diets might reduce the incidence of breast or colon cancer. These earlier studies were based on indirect evidence—for example, finding that women who moved to the United States from countries where diets were low in fat acquired a higher risk of cancer. There were also some animal studies showing that a high-fat diet could lead to more mammary cancer.

So the Women's Health Initiative was a serious attempt to directly test a central and very pressing question. A sign of the researchers' diligence can be seen in the surprisingly low rate of attrition. After eight years, only 4.7 percent of the women in the low-fat diet group withdrew from participation or were lost to follow-up (compared with 4.0 percent of the women in the regular diet group).

Nonetheless, the study has been attacked. Even supporters of evidence-based medicine have argued that the study wasted hundreds of millions of dollars because it asked the wrong question. Some say that the recommended diet wasn't low fat enough. The dieters were told that 20 percent of their calories could come from fat. (Only 31 percent of them got their dietary fat that low.) Some critics think—especially because of compliance issues—that the researchers should have recommended 10 percent fat.

Others critics think the study is useless because it tested only for the impact of reducing total fat in the diet instead of testing the impact of reducing saturated fats, which raise cholesterol levels. Randomized studies can't tell you anything about treatments that you failed to test. So we just don't know whether reducing saturated and trans fats might still reduce the risk of heart disease. And we're not likely to get the answer soon. Dr. Michael Thun, who directs epidemiological research for the American Cancer Society, called the WHI study, "the Rolls-Royce of studies," not just because it was high quality, but also because it was so expensive. "We usually have only one shot," he said, "at a very large-scale trial on a particular issue."

Similar concerns have been raised about another WHI study, which tested the impact of calcium supplements. A seven-year randomized test of 36,000 women aged fifty to seventy-nine found that taking calcium supplements resulted in no significant reduction in risk of hip fracture (but did increase the risk of kidney stones). Critics again worry that the study asked the wrong question to the wrong set of women. Proponents of calcium supplements want to know whether supplements might not still help older women. Others said they should have excluded women who are already getting plenty of calcium in their regular diet, so that the study would have tested the impact of calcium supplements when there is a pre-existing deficiency. And of course some wished they had tested a higher dose supplement.

Still, even the limited nature of the results gives one pause. Dr. Ethel Siris, president of the National Osteoporosis Foundation, said the new study made her question the advice she had given women to take calcium supplements regardless of what is in their diet. "We didn't think it hurt, which is why doctors routinely gave it," Siris said.

When she heard about the results of the calcium study, Siris's first reaction was to try to pick it apart. She changed her mind when she heard the unreasonable way that people were criticizing some of the WHI studies. Seeing the psychology of resistance in others helped her overcome it in herself. She didn't want to find herself "thinking there was something wrong with the design of this study because I don't like the results."

Much is at stake here. The massive randomized WHI studies are changing physician practice with regard to a host of treatments. Some doctors have stopped recommending low-fat diets to their patients as a way to reduce their heart disease and cancer risk. Others, like Siris, have changed their minds about calcium supplements. Even the best studies need to be interpreted. Done well, Super Crunching is a boon to society. Done badly, database decision making can kill.

The rise of Super Crunching is a phenomenon that cannot be ignored. On net, it has and will continue to improve our lives. Having more information about "what causes what" is usually good. But the purpose of this chapter has been to point out exceptions to this general tendency. Much of the resistance that we've seen over and over in this book can be explained by self-interest. Traditional experts don't like the loss of control and status that often accompanies a shift toward Super Crunching. But some of the resistance is more visceral. Some people fear numbers. For these people, Super Crunching is their worst nightmare. To them, the spread of data-driven decision making is just the kind of thing they thought they could avoid by majoring in the humanities and then studying something nice and verbal, like law.

We should expect a Super Crunching backlash. The greater its impact, the greater the resistance—at least pockets of resistance. Just as we have seen the rise of hormone-free milk and cruelty-free cosmetics, we should expect to see products that claim to be "data-mining free." In a sense, we already are. In politics, there is a certain attraction to candidates who are straight shooters, who don't poll every position, who don't relentlessly stay on message to follow a focus group–approved script. In business, we find companies like Southwest Airlines that charge one price for any seat on a particular route. Southwest passengers don't need Farecast to counter-crunch future fares on their behalf, because Southwest doesn't play the now-you-see-it-now-you-don't pricing games (euphemistically called "revenue enhancement") where other airlines try to squeeze as much as they can from every individual passenger.

While price resistance is reasonable, a broader quest for a life untouched by Super Crunching is both infeasible and ill-advised. Instead of a Luddite rejection of this powerful new technology, it is better to become a knowledgeable participant in the revolution. Instead of sticking your head in the sands of innumeracy, I recommend filling your head with the basic tools of Super Crunching.

CHAPTER 8

The Future of Intuition (and Expertise)

Here's a fable that happens to be true. Once upon a time, I went for a hike with my daughter Anna, who was eight years old at the time. Anna is a talkative girl who is, much to my consternation, developing a fashion sense. She's also an intricate planner. She'll start thinking about the theme and details of her birthday party half a year in advance. Recently, she's taken to designing and fabricating elaborate board games for her family to play.

While we were hiking, I asked Anna how many times in her life she had climbed the Sleeping Giant trail. Anna replied, "Six times." I then asked what was the standard deviation of her estimate. Anna replied, "Two times." Then she paused and said, "Daddy, I want to revise my mean to eight."

Something in Anna's reply gets at the heart of why "thinking-by-numbers is the new way to be smart." To understand what was going on in that little mind of hers, we have to step back and learn something about our friend, the standard deviation.

You see, Anna knows that standard deviations are an incredibly intuitive measure of dispersion. She knows that standard deviations give us a way of toggling back and forth between numbers and our intuitions about the underlying variability of some random process. This all sounds horribly abstract and unhelpful, but one concrete fact is now deeply ingrained in Anna's psyche:

> There's a 95 percent chance that a normally distributed variable will fall within two standard deviations (plus or minus) of its mean.

In our family, we call this the "Two Standard Deviation" rule (or 2SD for short). Understanding this simple rule is really at the heart of understanding variability. So what does it mean? Well, the average IQ score is 100 and the standard deviation is 15. So the 2SD rule tells us that 95 percent of people will have an IQ between 70 (which is 100 minus two standard deviations) and 130 (which is 100 plus two standard deviations). Using the 2SD rule gives us a simple way to translate a standard deviation number into an intuitive statement about variability. Because of the 2SD rule, we can think about variability in terms of something that we understand: probabilities and proportions. Most people (95 percent) have IQs between 70 and 130. If the distribution of IQs were less variable—say, the standard deviation was only 5—then the range of scores that just included 95 percent of the population would be much smaller. We'd be able to say 95 percent of people have IQs between 90 and 110. (In fact, later on, we'll learn how Larry Summers, the ousted president of Harvard, got into a world of trouble by suggesting that men and women have different IQ standard deviations.)

We now know enough to figure out what was going on in Anna's eight-year-old mind during that fateful hike. You see, Anna can recite the 2SD rule in her sleep. She knows that standard deviations are our friends and that the first thing you always do whenever you have a standard deviation and a mean is to apply the 2SD rule.

Recall that after Anna said she had hiked Sleeping Giant six times, she said the standard deviation of her estimate was two. She got the number two as her estimate for the standard deviation by thinking about the 2SD rule. Anna asked herself what the 95 percent range of her confidence was and then tried to back out a number that was consistent with her intuitions. She used the 2SD rule to translate her intuitions into a number. (If you want a challenge, see if you can use the 2SD rule and just your intuition to derive a number for the standard deviation for adult male height. You'll find help at the bottom of the page.)*

But Anna wasn't done. The really amazing thing was that after a pause of a few seconds, she said, "Daddy, I want to revise my mean to eight." During that pause, after she told me her estimate was six and the standard deviation was two, she was silently thinking more about the 2SD rule. The rule told her, of course, that there was a 95 percent chance that she had walked to the top of Sleeping Giant between two and ten times. And here's the important part: without any prompting she reflected on the truth of this range using nothing more than her experience, her memories. She realized that she had clearly walked it more than two times. Her numbers didn't fit her intuitions.

Anna could have resolved the contradiction by revising down her standard deviation estimate ("Daddy, I want to revise my standard deviation to one"). But she felt instead that it was more accurate to increase her estimate of the mean. By revising the mean up to eight, Anna was now saying there was a 95 percent chance that she had

* To use the 2SD rule, you need to estimate two things. First, what do you think is the average height of adult men in the U.S.? If you said 5'9", you're doing well. Now here's the harder question. Ninety-five percent of adult males fall between what two heights? Try to have your range of heights centered on the average male height of 5' 9". Forget about standard deviations and just answer the question based on what you know about the world.

The adult male height is distributed almost normally so that if your answer is correct, there will be 2.5 percent of men who are below your lower height and 2.5 percent of men who are taller than your upper height. Don't worry about being precise. Force yourself to write down a lower and an upper height. The answer is in the endnotes.

walked the trail between four and twelve times. Having been her companion on these walks, I can attest that she revised the right number.

I had never been more proud of my daughter. Yet the point of the story is not (only) to kvell about Anna's talents. (She is a smart kid, but she's no genius child. Notwithstanding my best attempts to twist her intellect, she's pretty normal.) No, the point of the story is to show how statistics and intuition can comfortably interact. Anna toggled back and forth between her memories as well as her knowledge of statistics to come up with a better estimate than either could have produced by themselves.

By estimating the 95 percent probability range, Anna actually produced a more accurate estimate of the mean. This is potentially a huge finding. Imagine what it could mean for the examination of trial witnesses, where lawyers often struggle to elicit estimates of when or how many times something occurred. You might even use it yourself when trying to jog your own or someone else's memory.

The (Wo)Man of the Future

For the rational study of the law the blackletter man may be the man of the present, but the man of the future is the man of statistics . . .

OLIVER WENDELL HOLMES, JR.

THE PATH OF THE LAW, 1897

The rise of statistical thinking does not mean the end of intuition or expertise. Rather, Anna's revision underscores how intuition will be reinvented to coexist with statistical thinking. Increasingly, decision makers will switch back and forth between their intuitions and data-based decision making. Their intuitions will guide them to ask new questions of the data that non-intuitive number crunchers would miss. And databases will increasingly al-low decision makers to test their intuitions—not just once, but on an ongoing basis.

This dialectic is a two-way street. The best data miners will sit back and use their intuitions and experiential expertise to query whether their statistical analysis makes sense. Statistical results that diverge widely from intuition should be carefully interrogated. While there is now great conflict between dyed-in-the-wool intuitivists and the new breed of number crunchers, the future is likely to show that these tools are complements more than substitutes. Each form of decision making can pragmatically counterbalance the greatest weaknesses of the other.

Sometimes, instead of starting with a hypothesis, Super Crunchers stumble across a puzzling result, a number that shouldn't be there. That's what happened to Australian economist Justin Wolfers, when he was teaching a seminar at the University of Pennsylvania's Wharton School on information markets and sports betting. Wolfers wanted to show his students how accurate Las Vegas bookies were at predicting college basketball games. So Wolfers pulled data on over 44,000 games—almost every college basketball game over a sixteen-year period. He created a simple graph showing what the actual margin of victory was relative to the market's predicted point spread.

"The graph was bang on a normal bell curve," he said. Almost ex-actly 50 percent (50.01 percent) of the time the favored team beat the point spread and almost exactly 50 percent of the time they came up short. "I wanted to show the class that this wasn't just true in general, but that it would hold true for different-size point spreads." The graphs that Wolfers made for games where the point spread was less than six points, and for point spreads from six to twelve points, again showed that the Las Vegas line was extremely accurate. Indeed, the following graph for all games where the point spread was twelve or less shows just how accurate:

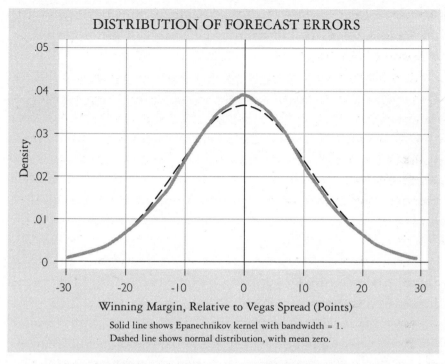

Solid line shows Epanechnikov kernel with bandwidth = 1.
Dashed line shows normal distribution, with mean zero.

SOURCE: Justin Wolfers, "Point Shaving: Corruption in NCAA Basketball,"
PowerPoint presentation, AEA Meetings (January 7, 2006)

Look how close the actual distribution of victory margins (the solid line) was to the theoretical normal bell curve. This picture gives you an idea of why they call it the "normal" distribution. Many real-world variables are approximately normal, taking the shape of your standard-issue bell curve. Almost nothing is perfectly normal. Still, many actual distributions are close enough to the normal distribution to provide a wickedly accurate approximation until you get several standard deviations into the tails.*

The problem was when Wolfers graphed games where the point

* By the way, the standard deviation for the margin of victory was 10.9 points. Can you apply the 2SD rule and say something intuitive about the variability of the actual victory margin relative to the Las Vegas line? It means that 95 percent of the time the actual margin of victory will be within about 21 points of the Las Vegas point spread.

spread was more than twelve points. When he crunched the numbers for his class, this is what he found:

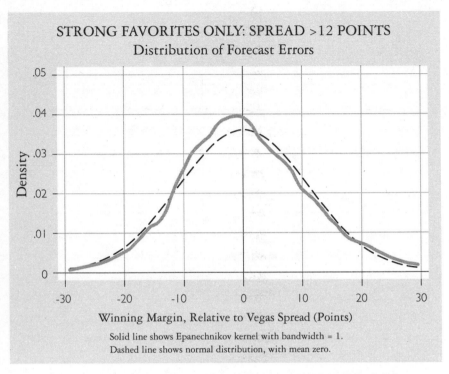

STRONG FAVORITES ONLY: SPREAD >12 POINTS
Distribution of Forecast Errors

Winning Margin, Relative to Vegas Spread (Points)

Solid line shows Epanechnikov kernel with bandwidth = 1.
Dashed line shows normal distribution, with mean zero.

SOURCE: Justin Wolfers, "Point Shaving: Corruption in NCAA Basketball," PowerPoint presentation, AEA Meetings. (January 7, 2006)

Instead of a 50–50 chance that the favored team would beat the point spread, Wolfers found there was only a 47 percent chance that the favored team would beat the spread (and hence a 53 percent chance that they would fail to cover the spread). This six percentage point difference might not sound like a lot, but when you're talking about millions of dollars bet on thousands of games (more than one fifth of college games have a point spread of more than twelve points), six percentage points is a big discrepancy. Something about the graph struck Wolfers as being very fishy and he started puzzling about it.

Wolfers was a natural for this investigation. He once worked for a bookie back home in Australia. More importantly, he is a leader in the new style of Super Crunching. A 2007 *New York Times* article, aptly titled "The Future of Economics Isn't So Dismal," singled him out as one of thirteen promising young economists, Super Crunchers all, who were remaking the field. Wolfers has a toothy smile and long, platinum blond hair often worn in a ponytail. To see him present a paper is to experience an endearing mixture of substance and flash. He is very much a rock star of Super Crunching.

So when Wolfers looked at the lopsided graph, it wasn't just that the favored team didn't cover the spread enough that bothered him; it was that they failed by just a few points. It was the hump in the distribution just below the Vegas line that didn't seem right. Justin began to worry that a small proportion of time when the point spread was high, players on the favored team would shave points. Suddenly, it all made a lot of sense. When there was a large point spread, players could shave points without really hurting their team's chances of still winning the game. Justin didn't think that all the games were rigged. But the pattern in the graph is what you'd see if about 6 percent of all high-point-spread games were fixed.

Justin didn't stop there. The future belongs to the Super Cruncher who can work back and forth and back again between his intuitions and numbers. The graph led Justin to hypothesize about point shaving. His hypothesizing led him to look for further tests that could confirm or disconfirm his hypothesis. He dug further in and found that if you looked at the score five minutes before the end of the game, there was no shortfall. The favored team was right on track to beat the spread 50 percent of the time. It was just in the last five minutes that the shortfall appeared. This isn't proof positive, but it does make a stronger circumstantial case—after all, that's the safest time for a bribed player to let up a bit, secure in the knowledge that it won't lead to his team losing.

The future belongs to people like Wolfers who are comfortable with both intuition and numbers. This "new way to be smart" is also for the consumers of Super Crunching. Increasingly, it will be useful for people like Anna to be able to quantify their intuitions. It is also important to be able to restate other people's Super Crunching results in terms that internally make intuitive sense.

One of the very coolest things about Super Crunching is that it not only predicts but also simultaneously tells you how accurate its prediction is. The standard deviation of the prediction is the crucial measure of accuracy. Indeed, the 2SD rule is the key to understanding whether a prediction is so accurate that Super Crunchers say it is "statistically significant." When statisticians say that a result is statistically significant, they are really just saying that some prediction is more than two standard deviations away from some other number. For example, when Wolfers says that the shortfall in favored teams covering the spread is statistically significant, he means that their 47 percent probability of covering is more than two standard deviations below the 50 percent probability he would predict if it was really a fair bet.

The designation of "statistical significance" is taken by a lot of people to be some highly technical determination. Yet it has a very intuitive explanation. There is less than a 5 percent chance that a random variable will be more than two standard deviations away from its expected mean (this is just the flip side of the 2SD rule). If an estimate is more than two standard deviations away from some other number, we say this is a statistically significant difference because it is highly unlikely (i.e., there is less than a 5 percent probability) that the estimated difference happened by chance. So just by knowing the 2SD rule, you know a lot about why "statistical significance" is pretty intuitive.

In this chapter, I hope to give you an idea for what it feels like to toggle back and forth between intuitions and numbers. I'm going to do it by introducing you to two valuable quantitative tools for the man or woman of the future. Reading this will not train you enough to be a full-fledged Super Cruncher. Yet learning and playing with these tools will put you well on the road to the wonderful dialectic of combining intuitions and statistics, experience and estimates. You've already started to learn how to use the first tool—the intuitive measure of dispersion, the standard deviation. One of the first steps is to see if you can communicate what you know to someone else.

A World of Information in a Single Number

When I taught at Stanford Law School, professors were required to award grades that had a 3.2 mean. Students would still obsess about how professors graded, but instead of focusing on the professor's mean grade, they'd obsess about how variable the grades were around the mandatory mean. Innumerable students and professors would engage in inane conversations where students would ask if a professor was a "spreader" or "clumper." Good students would want to avoid clumpers so that they would have a better chance at getting an A, while bad students hated the spreaders who handed out more As but also more Fs.

The problem was that many of the students and many of the professors had no way to express the degree of variability in professors' grading habits. And it's not just the legal community. As a nation, we lack a vocabulary of dispersion. We don't know how to express what we intuitively know about the variability of a distribution of numbers.

The 2SD rule could help give us this vocabulary. A professor who said that her standard deviation was .2 could have conveyed a lot of information with a single number. The problem is that very few people in the U.S. today understand what this means. But you should know and be able to explain to others that only about 2.5 percent of the professor's grades are above 3.6.

It's amazing the economy of information that can be conveyed in just a few words. We all know that investing in the stock market is risky, but just how risky is risky? Once again, standard deviations and the 2SD rule come to our rescue. Super Crunching regression tells us that the predicted return next year of a diversified portfolio of New York Stock Exchange stocks is 10 percent, but that the standard deviation is 20 percent. Just knowing these two numbers reveals an incredible amount.

Suddenly we know that there's a 95 percent chance that the return on this portfolio will be between minus 30 percent and positive 50 percent. If you invest $100, there's a 95 percent chance that you'll end the year with somewhere between $70 and $150. The actual returns on the stock market aren't perfectly normal, but they are close enough for us to learn an awful lot from just two numbers, the mean and the standard deviation.

Indeed, once you know the mean and standard deviation of a normal distribution, you know everything there is to know about the distribution. Statisticians call these two values "summary statistics" because they summarize all the information contained in the entire bell curve. Armed with a mean and a standard deviation, we can not only apply the 2SD rule; we can also figure out the chance that a variable will fall within any given range of values. Want to know the chance that the stock market will go down this coming year? Well, if the expected return is 10 percent and the standard deviation is 20 percent, you're really asking for the chance that the return will fall more than one half of a standard deviation below the mean. Turns out the answer (which takes about thirty seconds to calculate in Excel) is 31 percent.

Exploiting this ability to figure out the probability that some variable will be above or below a particular value pays even greater dividends in political polls.

Probabilistic Leader of the Pack

The current newspaper conventions on how to report polling data are all screwed up. Newspaper articles tend to say something like: "In a Quinnipiac poll of 1,243 likely voters, Calvin holds a 52 percent to 48 percent advantage over Hobbes for the Senate seat. The poll's margin of error is plus or minus two percentage points."

How many people understand what the margin of error really means? Do you? Before going on, write down what you think is the chance that most people in the state really support Calvin.

It should come as no surprise that the margin of error is related to the font of all statistical wisdom, the 2SD rule. The margin of error is nothing more than two standard deviations. So if the newspaper tells you that the margin of error is two percentage points, that means that one standard deviation is one percentage point. We want to know what proportion of people in the entire state population of likely voters support Calvin and Hobbes, but the sample proportions by chance might be misrepresentative of the population proportions. The standard deviation measure tells us how far the sample predictions might stray by chance from the true population proportions that we care about.

So once again we can apply our friend, the 2SD rule. We start with the sample proportion that supports Calvin, 52 percent, and then construct a range of numbers by adding on and subtracting off the margin of error (which is two standard deviations). That's 52 percent plus or minus 2 percent. So using the 2SD rule we can say, "There is a 95 percent chance that somewhere between 50 percent and 54 percent of likely voters support Calvin." Printing something like this would provide a lot more information than the cryptic margin of error disclaimer.

Even this 95 percent characterization fails, however, to emphasize an even more basic result: the probability that Calvin is actually leading. For this example, it's pretty easy to figure out. Since there is a 95 percent chance that Calvin's true support in the state is between 50 percent and 54 percent, there is a 5 percent chance that his true support is in one of the two tails of the bell curve—either above 54 percent or below 50 percent. And since the two tails of the bell curve are equal in size, there is just a 2.5 percent chance that Calvin's statewide support is less than 50 percent. That means there's about a 97.5 percent chance that Calvin is leading.

Reporters are massively misinformed when it comes to figuring out the probability of leading. If Laverne is leading Shirley 51 percent to 49 percent with a margin of error of 2 percent, news articles will say that the race is "a statistical dead heat." Balderdash, I say. Laverne's polling result is a full standard deviation above 50 percent. (Remember, the margin of error is two standard deviations, so in this example one standard deviation is 1 percent.) Crunching these numbers in Excel tells us in a few seconds that there is an 84 percent chance that Laverne currently leads in the polls. If something doesn't change, she is your likely winner.

In many polls, there are undecideds and third-party candidates, so the proportions of the two leading candidates often add up to less than 100 percent. But the probability of leading tells you just what it says—the probable leader of the pack.

People have a much easier time understanding proportions and probabilities than they do standard deviations and margins of error. The beauty of the 2SD rule is that it provides a bridge for translating one into the other. Instead of reporting the margin of error, reporters should start telling people something that they intuitively understand, the "probability of leading." Standard deviations are our friends, and they can be used to tell even the uninitiated about things that we really do care about.

Working Backwards

But wait, there's more. The stock and survey examples show that if you know the mean and standard deviation, you can work forward to calculate a proportion or probability that tells people something interesting about the underlying process. Yet sometimes it's useful to work backward, starting with a probability and then estimating the implicit standard deviation that would give rise to that result. Lawrence Summers got into a lot of trouble for doing just this.

On January 14, 2005, the president of Harvard University, Lawrence Summers, touched off a firestorm of criticism when he spoke at a conference on the scarcity of women professors in science and math. A slew of newspaper articles characterized his remarks as suggesting that women are "somehow innately deficient in mathematics." The *New York Times* in 2007 characterized Summers's remarks as claiming that "a lack of intrinsic aptitude could help explain why fewer women than men reach the top ranks of science and math in universities." The article (like many others) suggested that the subsequent furor over Summers's speech contributed to his resignation in 2006 (and the decision to replace him with the first female president in the university's 371-year history).

Summers's speech did in fact suggest that there might be innate differences in the intelligence of men and women. But he didn't argue that the average intelligence of women was any less than that of men. He focused instead on the possibility that the intelligence of men is more variable than that of women. He explicitly worked backwards from observed proportions to implicit standard deviations. Here's what Summers said:

> I did a very crude calculation, which I'm sure was wrong and certainly was unsubtle, twenty different ways. I looked . . . at the evidence on the sex ratios in the top 5 percent of twelfth graders [in science and math]. If you look at those—they're all over the map [but] one woman for every two men would be a high-end estimate [for the relative prevalence of women]. From that, you can back out a difference in the implied standard deviations that works out to be about 20 percent.

Summers doesn't say it, but his calculation assumes what researchers have in fact found: there is no pronounced difference in the *average* math or science scores for male and female twelfth graders. But in a variety of different studies, researchers have found a difference in the tails of the distribution. In particular, Summers focused in on the tendency for there to be two men for every one woman when you looked at the top 5 percent of math and science achievement among twelfth graders. Summers worked backwards to figure out what kind of a difference in standard deviations would give rise to this sex difference in the tails. His core claim, indeed his only claim, of innate difference was that the standard deviation of men's intelligence might be 20 percent greater than that of women.

Summers in the speech was careful to point out that his calculation was "crude" and "unsubtle." But Summers is no dummy. He is the youngest person ever to be voted tenure at Harvard. He won the prestigious John Bates Clark award for the best U.S. economist under forty. Two of the three greatest American economists of the twentieth century, Kenneth Arrow and Paul Samuelson, are his uncles, and like them Summers in his early forties was headed straight for a Nobel Prize. He definitely understands standard deviations. But after almost dying from Hodgkin's disease, Summers chose a different path. Like Paul Gertler of Progresa fame, he became chief economist for the World Bank and eventually went on to be secretary of the treasury at the end of the Clinton administration. He is almost always the smartest person in the room (and his critics say he knows it).

Being smart, however, does not mean that everything you say is right. Summers's back-of-the-envelope empiricism doesn't definitely resolve the question of whether women have less variable intelligence. For example, lots of other factors could have influenced the math and science scores of twelfth graders besides innate ability. Yet there have been subsequent studies suggesting that the IQ scores of women are in fact less variable than those of men.

Summers, in suggesting a gendered difference in standard deviations, is suggesting that men are more likely to be really smart, but he's also implying that men are innately more likely to be really dumb. It's a tricky question to know whether it is desirable to be associated with a group that has a larger IQ standard deviation. Imagine that you are expecting your first child. You are told that you can choose the range of possible IQs that your child will have but this range must be centered on an IQ of 100. Any IQ within the range that you choose is equally likely to occur. What range would you choose—95 to 105, or would you roll the dice on a wider range of, say, 60 to 140? When I asked this question to a group of fourth and sixth graders, they invariably chose ranges that were incredibly small (nothing wider than 95–105). None of them wanted to roll the dice on the chance that their kid would be a genius if it meant that their kid might alternatively end up as developmentally disabled. So from the kids' perspective, Summers was suggesting that men have a less desirable IQ distribution.

What really got Summers in trouble was taking his estimated 20 percent difference and using it to figure out other probabilities. Instead of looking at the ratio of males to females in the top 5 percent of the most intelligent people, he wanted to speculate about the ratio of men to women in the top 0.01 of 1 percent of the most scientifically intelligent people. Summers claimed that research scientists at top universities come from this more rarefied strata:

> If... one is talking about physicists at a top twenty-five research university, one is not talking about people who are two standard deviations above the mean. And perhaps it's not even talking about somebody who is three standard deviations above the mean. But it's talking about people who are three and a half, four standard deviations above the mean in the 1 in 5,000, 1 in 10,000 class.

To infer what he called the "available pool" of women and men this far out in the distribution, Summers took his estimates of the implicit standard deviations and worked forward:

> Even small differences in the standard deviation will translate into very large differences in the available pool substantially out [in the tail of the distribution].... [Y]ou can work out the difference out several standard deviations. If you do that calculation—and I have no reason to think that it couldn't be refined in a hundred ways—you get five to one, at the high end.

Summers was claiming that women may be underrepresented in science because for the kinds of smarts you need at a top research department, there might be five men for every one woman. Now you can start to understand why he got in so much trouble. I've recalculated

Summers's estimates, using his same methodology, and his bottom-line characterization of the results, if anything, was understated. At three point five or four standard deviations above the mean, a 20 percent difference in standard deviations can easily translate into there being ten or twenty times as many men as women. However, these results are far from definitive. I agree with him that his method might be flawed in "twenty different ways."

Still, I come to praise his process (if not the general conclusion he drew from it). Summers worked both backwards and forwards to derive a probability of interest. He started with an observed proportion, and with it he backed out implicit standard deviations for male and female scientific intelligence. He then worked forwards to estimate the relative number of women and men at another point in the distribution. This exercise didn't work out so well for Summers. Nonetheless, like Summers, the intuitivist of the future will keep an eye out for proportions and observed probabilities and use them to derive standard deviations (and vice versa).

The news media almost completely ignored the point that Summers was just talking about a difference in variability. It's not nearly as sexy as reporting "Harvard President Says Women Are Innately Deficient in Mathematics." (They might as easily have reported that Summers was claiming that women are innately superior in mathematics, since they are less likely to be really bad in math.) Many reporters simply didn't understand the point or couldn't figure out a way to communicate it to a general audience. It's a hard idea to get across to the uninitiated. At least in small part, Summers may have lost his job because people don't understand standard deviations.

This inability to speak to one another about dispersion hinders our ability to make decisions. If we can't communicate the probability of worst-case scenarios, it becomes a lot harder to take the right precautions. Our inability to communicate even impacts something as basic and important as how we plan for pregnancy.

Polak's Pregnancy Problems

Everybody knows that a baby is due roughly nine months after conception. However, few people know that the standard deviation is fifteen days. If you're pregnant and are planning to take off time from work or want to schedule a relative's visit, you might want to know something about the variability of when you'll actually give birth. Knowing the standard deviation is the best place to start. (The distribution is also skewed left—so that there are more pregnancies that are three weeks early than three weeks late.)

Most doctors don't even give the most accurate prediction of the due date. They still often calculate the due date based on the quasi-mystical formula of Franz Naegele, who believed in 1812 that "pregnancy lasted ten lunar months from the last menstrual period." It wasn't until the 1980s that Robert Mittendorf and his coauthors crunched numbers on thousands of births to let the numbers produce a formula for the twentieth century. Turns out that pregnancy for the average woman is eight days longer than the Naegele rule, but it's possible to make even more refined predictions. First-time mothers deliver about five days later than mothers who have already given birth. Whites tend to deliver later than nonwhites. The age of the mother, her weight, and her nutrition all help predict her due date.

Physicians using the crude Naegele rule cruelly set up first-time mothers for disappointment. These mothers are told that their due date is more than a week before their baby is really expected. And there is almost never any discussion about the probable variability of the actual delivery.

So right off the bat, physicians are failing to communicate. As Ben Polak quickly found out, it gets worse. Ben, in his day job, is a theoretical economist. In the typically rumpled raiment of academics, he prowls his seminar rooms with piercing eyes. Polak was born in London and has a proper British accent (when I was reading *Jane Eyre,* he politely enlightened me as to the correct pronunciation of "St. John").

On the side, Ben crunches numbers in the fine tradition of Bill James. Indeed, Ben and his coauthor Brian Lonergan have done James one better. Instead of predicting the contribution of baseball players to scoring runs, Polak made a splash when he estimated each player's contributions to his team's wins—the true competitive bottom line. There is a satisfying simplicity to Polak's estimates because the players' contributions to a team sum up to the number of games the team actually wins.

Ben wasn't nearly as satisfied about the statistics he was given by health care professionals when his wife, Stefanie, was pregnant with their first child, Nelly. And the problem was a lot more important than predicting their due date. Ben and Stefanie wanted to know the chance that their child would have Down syndrome.

I remember when my partner, Jennifer, and I were expecting for the first time—back in 1994. Back then, women were first told the probability of Down syndrome based on their age. After sixteen weeks, the mother could have a blood test measuring her alphafetoprotein (AFP) level, and then they'd give you another probability. I remember asking the doctor if they had a way of combining the different probabilities. He told me flat out, "That's impossible. You just can't combine probabilities like that."

I bit my tongue, but I knew he was dead wrong. It is possible to combine different pieces of evidence, and has been since 1763 when a short essay by the Reverend Thomas Bayes was posthumously published. Bayes' theorem is really a single equation that tells us how to update an initial probability given a new piece of evidence.

Here's how it works. A woman who is thirty-seven has about a 1 in 250 (.4 percent) chance of giving birth to a child with Down syndrome. The Bayes formula tells you how you can update this probability to take into account a woman's AFP level. To update, all you need to do is multiply the initial probability by a single number, a "likelihood ratio" which might either inflate or deflate the original probability estimate.*

Knowing the best single estimate from these early tests is important because some parents will want to go further and have amniocentesis if the probability of Down syndrome is too high. "Amniocentesis is almost 100 percent accurate," Polak said. "But it carries a risk. The risk is that the amniocentesis might cause a miscarriage of the child." Once in about every 250 procedures, a pregnant woman will miscarry after amnio.

There is some good news to report here. In the last decade, doctors have uncovered multiple predictors for Down syndrome. Instead of the single AFP test, there is now the triple screen, which uses Bayesian updating to predict the probability of Down syndrome based on three different assays from a single blood test. Doctors have also noticed that fetuses with Down syndrome are more likely to display on sonograms a thick patch of skin at the base of the neck—the so-called nuchal fold.

Nonetheless, Ben (like Bono) still hasn't found what he's looking for. He told me that the health care professionals would tend to downplay the importance of numbers. "I had a comical interaction with a very nice physician," he told me, "where she said, 'One of these probabilities is 1 in 1,000, and one is 1 in 10,000, so what's the difference?' and to me there's actually quite a big difference. There's a ten times difference." I had a similar interaction with a genetic counselor in California after Jennifer's blood showed a risky level of AFP. When I asked for a probability of Down syndrome, the counselor unhelpfully offered: "That would only be a number. In reality, your child will either have it or not."

In Ben's case, the health care providers suggested that Ben and

---

* The likelihood ratio measures how likely it is that you would have seen this AFP level if the baby in fact had Down syndrome. Technically, the likelihood ratio is the probability that a mother with a Down syndrome child would have this AFP score divided by the probability that the mother would have this AFP level (regardless of whether her baby has Down syndrome).

Stefanie follow a rule of thumb that to him was "completely arbitrary." He said:

> The rule that they guide you towards is to take amniocentesis when the probability of miscarriage is less than the probability of Down syndrome. That assumes that the parents put equal weight on the two bad outcomes. The two bad outcomes are miscarriage and Down syndrome. And there may be parents for whom those aren't at all equal. In fact, it could go either way.

Some parents might find the loss from a miscarriage to be much greater (especially, for example, if the couple is unlikely to be able to get pregnant again). However, as Ben points out, "[There are a] lot of parents for whom having a severely handicapped child would be devastating. And they would choose to put a big weight on that as a bad outcome."

One of the great things about Super Crunching is that it tends to tie decision rules to the actual consequences of the decision. More enlightened genetic counselors would do well to ask patients which adverse outcome (Down syndrome or miscarriage) they would experience as a greater loss—and how much greater a loss it would be. If a pregnant woman says that she would weigh Down syndrome three times worse than a miscarriage, then the woman should probably have amniocentesis if the probability of Down syndrome is more than a third greater than the probability of miscarriage.

Ben was also frustrated by what doctors told him about the triple screen results. "When you take the actual test, all you really want to know is a simple thing: what is the probability that my child has Down syndrome, given the outcome of the test. You don't want to know a bunch of extraneous information about what the probability of a false positive is here." The real power of the Bayes equation is that it gives you the bottom line probability of Down syndrome while taking into account the possibility that a test will have false positives.

Overall, there are now five valid predictors of Down syndrome—the mother's age, three blood tests, and the nuchal fold—that could help mothers and fathers decide whether to have amniocentesis. Yet even today, many doctors don't provide a bottom-line prediction that is based on all five factors. Ben said, "I had trouble getting them to combine the data from these blood and nuchal fold tests. There must be vast amounts of data to do this. It cannot be a hard exercise to do. But they just didn't have this analysis available. When I asked, the physician said something technical about one of the distributions being non-Gaussian. But that was utterly irrelevant to the issue at hand."

The Bayesian cup is more than half full. The quad screen now combines the information from four different predictors into a single, bottom-line probability of Down syndrome. Medical professionals, however, are still making "you can't get there from here" claims that full updating on the nuchal fold information just isn't possible.

The Bayes equation is the science of learning. It is the second great tool of this chapter. If the Super Cruncher of the future is really going to dialectically toggle back and forth between her intuitions and her statistical predictions, she's going to have to know how to update her predictions and intuitions over time as she gets new information. Bayes' equation is crucial to this updating process.

Still, it's not surprising that many health care professionals aren't comfortable with updating. We learned before that physicians have been good at biochemistry, but they often are still out to lunch when it comes to basic statistics. For example, several studies have asked physicians over the years the following type of question:

> One percent of women at age forty who participate in routine screening have breast cancer. Eighty percent of women with breast cancer will get positive mammographies. Ten percent of women without breast cancer will also get positive mammographies. A woman in this age group had a positive mammography in a routine screening. What is the probability that she actually has breast cancer?

By the way, this is a question you can answer, too. What do you think is the probability that a woman who has a positive mammography has breast cancer? Think about it for a minute.

In study after study, most physicians tend to estimate that the probability of cancer is about 75 percent. Actually, this answer is about ten times too high. Most physicians don't know how to apply Bayes' equation.

We can actually work out the probability (and learn Bayes to boot) if we translate the probabilities into frequencies. First, imagine a sample of 1,000 women who are screened for breast cancer. From the 1 percent (prior) probability, we know that 10 out of every 1,000 women who get screened will actually have breast cancer. Of these 10 women with breast cancer, 8 will have a positive mammogram. We also know that of the 990 women without breast cancer who take the test, 99 will have a false positive result. Can you figure out the probability that a woman with a positive test will have breast cancer now?

It's a pretty straightforward calculation. Eight out of the 107 positive tests (8 true positives plus 99 false positives) will actually have cancer. So what statisticians call the posterior or updated probability of cancer conditioned upon a positive mammogram becomes 7.5 percent (8 divided by 107). Bayes' theorem tells us that the prior 1 percent probability of cancer doesn't jump to 70 or 75 percent—it increases to 7.5 percent.

People who don't understand Bayes tend to put too much emphasis on the 80 percent chance that a woman with cancer will test positive. Most physicians studied seem to think that if 80 percent of women with breast cancer have positive mammographies, then the probability of a woman with a positive mammography having breast cancer must be around 80 percent. But Bayes' equation tells us why this intuition is wrong. We have to put a lot more weight on the original, unconditional fraction of women with breast cancer (the prior probability), as well as the possibility that women without breast cancer will receive false positives.

Can you figure out what the probability of cancer is for the women whose mammogram comes back negative? If you can (and the answer is provided below),* you're well on your way to mastering the updating idea.

When All Is Said and Done

Knowing about the 2SD rule and Bayes' theorem can improve the quality of your own decisions. Yet there are many more tools that you would need to master to become a bona fide Super Cruncher or even a reasonable consumer of Super Crunching. You'd need to become comfortable with terms like heteroskedasticity and omitted variable bias. This book isn't an end, it's an invitation. If you're hooked, the endnotes contain suggested readings for the future.

Like me, Ben Polak is passionate about the need to inculcate a basic understanding of statistics in the general public. "We have to get students to learn this stuff," he says. "We have to get over this phobia and we have to get over this view that somehow statistics is illiberal. There is this crazy view out there that statistics are right-wing." The stories in this book refute the idea that Super Crunching is part of some flattening right-wing conspiracy (or any other ideological hegemony). Super Crunching is what gives the Poverty Action Lab their persuasive power to improve the world. One can crunch numbers and still have a passionate and caring soul. You can still be creative. You just have to be willing to put your creativity and your passions to the test to see if they really work.

I've been speculating about how in the future intuition, expertise, and data-based analysis will interact. We'll see proliferation and deep resistance. We'll see both progress and advantage-taking. But the seeds of the future can be seen in the past. These are the themes that

---

* Two out of the 893 negative tests will have cancer—so the posterior probability falls to just .2 percent (one-fifth of the prior probability).

we've already seen play out time and again in the book. They were also present way back in 1957, in one of the lesser-known Katharine Hepburn–Spencer Tracy movies, *Desk Set*.

In the movie, Bunny Watson (played by Hepburn) is the super-smart head of a reference library for a large TV network. She's responsible for researching and answering questions on all manner of topics, such as the names of Santa's reindeer. Onto the scene comes Richard Sumner (played by Spencer Tracy), the Super Crunching inventor of the EMERAC computer, which Sumner affectionately nicknames "Emmy." The movie pits Bunny's encyclopedic memory against the immense "electronic brain," playing on the same kind of fears that we see today—the fear that in an increasingly automated world, traditional expertise will become irrelevant. Bunny and others are worried that they'll lose their jobs.

It's useful to reflect on just how lopsided the Bunny/Emmy competition has become. We now take for granted that computers are better at retrieving bits of information. Indeed, one need look no further than Google to realize that this is a fight which no human researcher has a prayer of winning. The Wikipedia page for the movie lists all the informational challenges that appeared in the movie itself with links to open source answers. Need to know the third stanza of the poem "Curfew Must Not Ring Tonight," by Rose Hartwick Thorpe? We all know the fastest way to find the answer (and it's not by phoning a friend).

I predict that we will increasingly come to look at the prediction competition between experts and Super Crunching equations in much the same way. It will just be accepted that it's not a fair fight. Super Crunching computers are much better than humans at figuring out what predictive weights to put on particular causal factors. When there are enough data, Super Crunching is going to win.

But *Desk Set* is also instructive in the way that it resolves the technological tension. Hepburn's character is not as quick at data retrieval as the EMERAC. In the end, however, the computer doesn't render people like Bunny useless. Her usefulness just changes, and the computer ultimately helps make her and other reference librarians more effective and efficient. The moral is that computers are tools that will make life easier and more fun. The movie isn't very subtle about this point and, indeed, right after the credits, there is a message about how helpful IBM was in making the movie.

Subtle or not, I think the same point applies to the rise of Super Crunching. In the end, Super Crunching is not a substitute for intuition, but a complement. This new way to be smart will not relegate humans to the trash heap of history. I'm not quite as sanguine about the future of traditional expertise. You don't need to watch the movie to know that technological dinosaurs who eschew the web are at a severe disadvantage in trying to retrieve information. The same thing goes for experts who resist the siren songs of Super Crunched predictions. The future belongs to those who can comfortably inhabit both worlds.

Our intuitions, our experiences, and, yes, statistics should work together to produce better choices. Of course, intuitions and experiential rules of thumb will still drive many of our day-to-day decisions. I doubt that we will see quantitative studies on the best way to fry an egg or peel a banana. Still, the experiences of thousands of other similarly situated people reduced to analyzable numbers can help us in ways that will be increasingly hard to ignore.

ACKNOWLEDGMENTS

The book has only my name on the spine. But let me raise a series of toasts to its many, many co-authors:

To Joyce Finan, my high-school math teacher, who told me that I'd never be any good at numbers because my handwriting was too messy.

To Jerry Hausman, my MIT econometrics professor, who taught me that there are some proofs that aren't worth knowing.

To Bob Bennett, Bill Felstiner, and the American Bar Foundation, who helped fund my first Super Crunching tests of car negotiations.

To my full-time data-crunching assistants, Fred Vars, Nasser Zakariya, Heidee Stoller, and most recently, Isra Bhatty. These trusting souls signed on for a year of nonstop crunching— usually on a dozen or more projects that were progressing simultaneously. Arlo Guthrie once said that it was hard for him to keep up his guitar skills when he had such talented side musicians. I know how he feels.

To Orley Ashenfelter, Judy Chevalier, Dick Copaken, Esther Duflo, Zig Engelmann, Paul Gertler, Dean Karlin, Larry Katz, Steve Levitt, Jennifer Ruger, Lisa Sanders, Nina Sassoon, Jody Sindelar, Petra Todd, Joel Waldfogel, and the many other people who gave of their time to make this book better.

To my research assistants at Yale Law School, Rebecca Kelly, Adam Banks, and Adam Goldfarb, who with unreasonable energy and care read and reread every word in this book.

To Lynn Chu and Glen Hartley, my agents, who beat me about the head until I had a reasonable proposal. Thank you for not giving up on me.

To John Flicker, my editor, who knows the value of carrots and sticks. Rarely has my writing improved so much after the initial draft, and you are the reason why.

To Peter Siegelman and John Donohue, to whom this book is dedicated. My memory of our years in Chicago still burns bright.

To Jennifer Brown, my best friend, who has sat with me in the wee hours of the morning as these pages grew into a manuscript.

And finally, to my other coauthors, Bruce Ackerman, Barry E. Adler, Antonia Ayres-Brown, Henry Ayres-Brown, Katharine Baker, Joe Bankman, John Braithwaite, Richard Brooks (who also gave detailed comments on the manuscript), Jeremy Bulow, Stephen Choi, Peter Cramton, Arnold Diethelm, Laura Dooley, Aaron Edlin, Sydney Foster, Matthew Funk, Robert Gaston, Robert Gertner, Paul M. Goldbart, Gregory Klass, Paul Klemperer, Sergey I. Knysh, Steven D. Levitt, Jonathan Macey, Kristin Madison, F. Clayton Miller, Edward J. Murphy, Barry Nalebuff, Eric Rasmussen, Stephen F. Ross, Colin Rowat, Peter Schuck, Stewart Schwab, Richard E. Speidel, and Eric Talley, who through the years have provided me with both intellectual and spiritual succor. Because of you, this is one gearhead who has found that a life with numbers and passion really can mix.

NOTES

INTRODUCTION

Pages 1–6: **Ashenfelter vs. Parker:** Stephanie Booth, "Princeton Economist Judges Wine by the Numbers: Ashenfelter's Analyses in 'Liquid Assets' Rarely off the Mark," *Princeton Packet*, Apr. 14, 1998, http://www.pac pubserver.com/new/news/4-14-98/wine.html; Andrew Cassel, "An Economic Vintage that Grows on the Vine," *Phila. Inquirer*, Jul. 23, 2006; Jay Palmer, "A Return Visit to Earlier Stories: Magnifique! The Latest Bordeaux Vintage Could Inspire Joyous Toasts," *Barron's*, Dec. 15, 1997, p. 14; Jay Palmer, "Grape Expectations: Why a Professor Feels His Computer Predicts Wine Quality Better than All the Tasters in France," *Barron's*, Dec. 30, 1996, p. 17; Peter Passell, "Wine Equation Puts Some Noses Out of Joint," *N.Y. Times*, Mar. 4, 1990, p. A1; Marcus Strauss, "The Grapes of Math," *Discover Magazine*, Jan. 1991, p. 50; Lettie Teague, "Is Global Warming Good for Wine?" *Food and Wine*, Mar. 2006, http://www.food andwine.com/articles/is-global-warming-good-for-wine; Thane Peterson, "The Winemaker and the Weatherman,"

Bus. Wk. Online, May 28, 2002, http://www.businessweek.com/bwdaily/dnflash/may2002/nf20020528_2081.htm.

7–9: **Bill James and baseball expertise:** Michael Lewis, *Moneyball: The Art of Winning an Unfair Game* (2003). For James's most recent statistical collection, see Bill James, *The Bill James Handbook 2007* (2006).

9: **Brown debuts for A's:** official site of Oakland A's player information, http://oakland.athletics.mlb.com/team/player-career.jsp?player_id=425852.

11: **On Kasparov and Deep Blue:** Murray Campbell, "Knowledge Discovery in Deep Blue: A Vast Database of Human Experience Can't Be Used to Direct a Search," 42 *Comm. ACM* 65 (1999).

12: **Testing which policies work:** Daniel C. Esty and Reece Rushing, *Data-Driven Policymaking*, Center for American Progress (Dec. 2005).

14–16: **The impact of LoJack:** Ian Ayres and Steven D. Levitt, "Measuring the Positive Externalities from Unobservable Victim Precaution: An Empirical Analysis of LoJack," 113 *Q. J. Econ.* 43 (1998).

CHAPTER 1

Pages 19–20: **Preference engine problems:** Alex Pham and Jon Healey, "Telling You What You Like: 'Preference Engines' Track Consumers' Choices Online and Suggest Other Things to Try. But Do They Broaden Tastes or Narrow Them?" *L.A. Times*, Sep. 20, 2005, p. A1; Laurie J. Flynn, "Amazon Says Technology, Not Ideology, Skewed Results," *N.Y. Times*, Mar. 20, 2006, p. 8; Laurie J. Flynn, "Like This? You'll Hate That. (Not All Web Recommendations Are Welcome.)," *N.Y. Times*, Jan. 23, 2006, p. C1; Ylan Q. Mui, "Wal-Mart Blames Web Site Incident on Employee's Error," *Wash. Post*, Jan. 7, 2006, p. D1.

21–22: **The exploitation of preference distributions:** Chris Anderson, *The Long Tail: Why the Future of Business Is Selling Less of More* (2006); Cass Sunstein, *Republic.com* (2001); Nicholas Negroponte, *Being Digital* (1995); Carl S. Kaplan, "Law Professor Sees Hazard in Personalized News," *N.Y. Times*, Apr. 13, 2001; Cass R. Sunstein, "Boycott the Daily Me!: Yes, the Net Is Empowering. But It Also Encourages Extremism—and That's Bad for Democracy," *Time*, Jun. 4, 2001, p. 84; Cass R. Sunstein, "The Daily We: Is the Internet Really a Blessing for Democracy?" *Boston Rev.*, Summer 2001.

22: **Accuracy of crowd predictions:** James Surowiecki, *The Wisdom of Crowds* (2004); Michael S. Hopkins, "Smarter Than You," Inc.com, Sep. 2005, http://www.inc.com/magazine/20050901/mhopkins.html.

22–28: **Internet dating loves Super Crunching:** Steve Carter and Chadwick

Snow, eHarmony.com, "Helping Singles Enter Better Marriages Using Predictive Models of Marital Success," Presentation to 16th Annual Convention of the American Psychological Society (May 2004), http://static.eharmony.com/images/eHarmony-APS-handout.pdf; Jennifer Hahn, "Love Machines," AlterNet, Feb. 23, 2005; Rebecca Traister, "My Date with Mr. eHarmony," Salon.com, Jun. 10, 2005, http://dir.salon.com/story/mwt/feature/2005/06/10/warren/index.html; "Dr. Warren's Lonely Hearts Club: EHarmony Sheds its Mom-and-Pop Structure, Setting the Stage for an IPO," *Bus. Wk.*, Feb. 20, 2006; Press Release, eHarmony.com, "Over 90 Singles Marry Every Day on Average at eHarmony," Jan. 30, 2006, http://www.eharmonyreviews.com/news2.html; Garth Sundem, *Geek Logik: 50 Foolproof Equations for Everyday Life* (2006).

26: **Prohibition of race discrimination in contracting:** Civil Rights Act of 1866, codified as amended at 42 U.S.C. 1981 (2000).

28: **Super Crunching in employment decisions:** Barbara Ehrenreich, *Nickel and Dimed: On (Not) Getting By in America* (2001).

29: **Companies use Super Crunching to maximize profitability:** Thomas H. Davenport, "Competing on Analytics," *Harv. Bus. Rev.*, Jan. 2006; telephone interview with Scott Gnau, Teradata vice president and general manager, Oct. 18, 2006.

29–32: **Corporate defection detection:** Keith Ferrell, "Landing the Right Data: A Top-Flight CRM Program Ensures the Customer Is King at Continental Airlines," *Teradata Magazine*, June 2005; "Continental Airlines Case Study: 'Worst to First,'" Teradata White Paper (2005), http://www.teradata.com/t/page/133201/index.html; Deborah J. Smith, "Harrah's CRM Leaves Nothing to Chance," *Teradata Magazine*, Spring 2001; Mike Freeman, "Data Company Helps Wal-Mart, Casinos, Airlines Analyze Customers," *San Diego Union Trib.*, Feb. 24, 2006.

32–34: **Ideas for increasing consumer information:** Barry Nalebuff and Ian Ayres, *Why Not?: How to Use Everyday Ingenuity to Solve Problems Big and Small* (2003); Peter Schuck and Ian Ayres, "Car Buying, Made Simpler," *N.Y. Times*, Apr. 13, 1997, p. F12.

34–37: **Farecast provides Super Crunching services to the consumer:** Damon Darlin, "Airfares Made Easy (or Easier)," *N.Y. Times*, Jul. 1, 2006, p. C1; Bruce Mohl, "While Other Sites List Airfares, Newcomer Forecasts Where They're Headed," *Boston Globe*, Jun. 4, 2006, p. D1; telephone interview with Henry Harteveldt, vice president and principal analyst at Forrester Research, Oct. 6, 2006; Bob Tedeschi, "An Insurance Policy for Low Airfares," *N.Y. Times*, Jan. 22, 2007, p. C10.

36: **Home price predictions:** Marilyn Lewis, "Putting Home-Value Tools to the

Test," MSN Money, www.moneycentral.msn.com/content/Banking/Home financing/P150627.asp.

36–37: **Accenture price predictions:** Daniel Thomas, "Accenture Helps Predict the Unpredictable," *Fin. Times*, Jan. 24, 2006, p. 6; Rayid Ghani and Hillery Simmons, "Predicting the End-Price of Online Auctions," ECML Workshop Paper (2004), available at http://www.accenture.com/NR/exeres/FO469E82-E904-4419-B34F-88D4BA53E88E.htm; telephone interview with Rayid Ghani, Accenture Labs researcher, Oct. 12, 2006.

41: **Catching a cell phone thief:** Ian Ayres, "Marketplace Radio Commentary: Cellphone Sleuth," Aug. 20, 2004; Ian Ayres and Barry Nalebuff, "Stop Thief!" *Forbes*, Jan. 10, 2005, p. 88.

41: **Counterterrorism social network analyis:** Patrick Radden Keefe, "Can Network Theory Thwart Terrorists?" *N.Y. Times Magazine*, Mar. 13, 2006, p. 16; Valdis Krebs, "Social Network Analysis of the 9-11 Terrorist Network," 2006, http://orgnet.com/hijackers.html; "Cellular Phone Had Key Role," *N.Y. Times*, Aug. 16, 1993, p. C11.

42: **Beating the magic number scams:** Allan T. Ingraham, "A Test for Collusion Between a Bidder and an Auctioneer in Sealed-Bid Auctions," 4 *Contributions to Econ. Analysis and Pol'y*, Article 10 (2005).

CHAPTER 2

Page 46: **Fisher proposes randomization:** Ronald Fisher, *Statistical Methods for Research Workers* (1925); Ronald Fisher, *The Design of Experiments* (1935).

47: **CapOne Super Crunching:** Charles Fishman, "This Is a Marketing Revolution," 24 *Fast Company* 204 (1999), www.fastcompany.com/online/24/capone.html.

50: **Point shaving in college basketball:** Justin Wolfers, "Point Shaving: Corruption in NCAA Basketball," 96 *Am. Econ. Rev.* 279 (2006); David Leonhardt, "Sad Suspicions About Scores in Basketball," *N.Y. Times*, Mar. 8, 2006.

50–52: **Other Super Crunching companies:** Haiyan Shui and Lawrence M. Ausubel, "Time Inconsistency in the Credit Card Market," 14th Ann. Utah Winter Fin. Conf. (May 3, 2004), http://ssrn.com/abstract=586622; Marianne Bertrand et al., "What's Psychology Worth? A Field Experiment in the Consumer Credit Market," Nat'l Bureau of Econ. Research, Working Paper No. 11892 (2005).

53–54: **Monster.com randomizes:** "Monster.com Scores Millions: Testing Increases Revenue Per Visitor," Offermatica, http://www.offermatica.com/stories-1.7.html (last visited Mar. 1, 2007).

54: **Randomizing ads for Jo-Ann Fabrics:** "Selling by Design: How Joann.com Increased Category Conversions by 30% and Average Order Value by 137%," Offermatica, http://www.offermatica.com/learnmore-1.2.5.2.html (last visited Mar. 1, 2007).

54: **Randomization goes non-profit:** Dean Karlan and John A. List, "Does Price Matter in Charitable Giving? Evidence from a Large-Scale Natural Field Experiment," Yale Econ. Applications and Pol'y Discussion Paper No. 13, 2006, http://ssrn.com/abstract=903817; Armin Falk, "Charitable Giving as a Gift Exchange: Evidence from a Field Experiment," CEPR Discussion Paper No. 4189, 2004, http://ssrn.com/abstract=502021.

58–59: **Continental copes with transportation events:** Keith Ferrell, "Teradata QandA: Continental—Landing the Right Data," *Teradata Magazine*, Jun. 2005, http://www.teradata.com/t/page/133723/index.html.

59: **Amazon apologizes:** Press release, Amazon.com, "Amazon.com Issues Statement Regarding Random Price Testing," Sep. 27, 2000, http://phx. corporate-ir.net/phoenix.zhtml?c=97664andp=irol-newsArticle_PrintandID= 229620andhighlight=.

CHAPTER 3

Page 64: **The $5 million thesis:** David Greenberg et al., *Social Experimentation and Public Policy-making* (2003). For more on income maintenance experiments, see Gary Burtless, "The Work Response to a Guaranteed Income: A Survey of Experimental Evidence 28," in *Lessons from the Income Maintenance Experiments*, Alicia H. Munnell, ed. (1986); Government Accountability Office, Rep. No. HRD 81-46, "Income Maintenance Experiments: Need to Summarize Results and Communicate Lessons Learned" 15 (1981); Joseph P. Newhouse, *Free for All?: Lessons from the Rand Health Insurance Experiment* (1993); Family Support Act of 1988, Pub. L. No. 100–485, 102 Stat. 2343 (1988).

66: **Super Crunching in the code and the UI tests:** Omnibus Budget Reconciliation Act of 1989, Pub. L. No. 101–239, § 8015, 103 Stat. 2470 (1989); Bruce D. Meyer, "Lessons From the U.S. Unemployment Insurance Experiments," 33 *J. Econ. Literature* 91 (1995).

66: **Search-assistance regressions:** Peter H. Schuck and Richard J. Zeckhauser, *Targeting in Social Programs: Avoiding Bad Bets, Removing Bad Apples* (2006).

66–67: **Alternatives to job-search assistance:** Instead of job-search assistance, other states tested whether reemployment bonuses could be effective in shortening the period of unemployment. These bonuses were essentially bribes for people to find work faster. A random group of the unemployed would be paid between $500 and $1,500 (between three and six times the weekly UI benefit) if they could find a job fast. The reemployment bonuses, however, were not generally successful in reducing the government's overall UI expenditures. The amount spent on the bonuses and administering the program often was larger than the amount saved in shorter unemployment spells. Illinois also tested whether it would be more effective to give the bonus to the employer or to the employee. Marcus Stanley et al., *Developing Skills: What We Know About the Impacts of American Employment and Training Programs on Employment, Earnings, and Educational Outcomes*, October 1998 (working paper).

68: **Good control groups needed:** Susan Rose-Ackerman, "Risk Taking and Reelection: Does Federalism Promote Innovation?" 9 *J. Legal Stud.* 593 (1980).

68: **Other randomized studies that are impacting real-world decisions:** The randomized trial has been especially effective at taking on the most intransigent and entrenched problems, like cocaine addiction. A series of randomized trials has shown that paying cocaine addicts to show up for drug treatment increases the chance that they'll stay clean. Offering the addicts lotteries is an even cheaper way to induce the same result. Rather than paying addicts a fixed amount for staying clean, participants in the lottery studies (who showed up and provided a clean urine sample) earned the chance to draw a slip of paper from a bowl. The paper would tell the addict the size of a prize varying from $1 to $100 (a randomized test about a randomized lottery). See Todd A. Olmstead et al., "Cost-Effectiveness of Prize-Based Incentives for Stimulant Abusers in Outpatient Psychosocial Treatment Programs," *Drug and Alcohol Dependence* (2006), http://dx.doi.org/10.1016/j.drugalcdep.2006.08.012.

68–69: **MTO testing:** Jeffrey R. Kling et al., "Experimental Analysis of Neighborhood Effects," *Econometrica* 75.1 (2007).

69: **Examples of researchers utilizing such tests to answer these questions:** Alan S. Gerber and Donald P. Green, "The Effects of Canvassing, Direct Mail, and Telephone Contact on Voter Turnout: A Field Experiment," 94 *Am. Pol. Sci. Rev.* 653 (2000); Yan Chen et al., "Online Fund-Raising Mechanisms: A Field Experiment," 5 *Contributions to Econ. Analysis and Pol'y* (2006), http://www.bepress.com/bejeap/contributions/vol5/iss2/art4; Stephen Ansolabehere and Shanto Iyengar, *Going Negative: How Political Advertisements Shrink and Polarize the Electorate* (1995).

69: **Testing college roommates:** Michael Kremer and Dan M. Levy, "Peer Effects and Alcohol Use Among College Students," Nat'l Bureau of Econ. Research Working Paper No. 9876 (2003) (men who drank prior to college earned much lower grades if they were placed with a heavy drinker instead of a teetotaler).

70: **Testing ballot order:** Daniel E. Ho and Kosuke Imai, "The Impact of Partisan

Electoral Regulation: Ballot Effects from the California Alphabet Lottery, 1978–2002," Princeton L. and Pub. Affairs Paper No. 04–001, Harvard Pub. L. Working Paper No. 89 (2004).

70: **For more information on Joel's work, see the following smorgasbord of Waldfogel empiricism:** Joseph Tracey and Joel Waldfogel, "The Best Business Schools: A Market-Based Approach," 70 *J. Bus.* 1 (1997); Joel Waldfogel, "The Deadweight Loss of Christmas: Reply," 88 *Am. Econ. Rev.* 1358 (1998); Felix Oberholzer-Gee et al., "Social Learning and Coordination in High-Stakes Games: Evidence from Friend or Foe," Nat'l Bureau of Econ. Research, Working Paper No. 9805 (2003), http://www.nber.org/papers/W9805; Joel Waldfogel, "Aggregate Inter-Judge Disparity in Federal Sentencing: Evidence from Three Districts," *Fed. Sent'g Rep.*, Nov.–Dec. 1991.

72: **Other market reactions:** Joel and I have also ranked judges' bail decisions based on how the bail bonds market reacted to the decisions. Ian Ayres and Joel Waldfogel, "A Market Test for Race Discrimination in Bail Setting," 46 *Stan. L. Rev.* 987 (1994).

72: **Post-conviction earnings:** Jeffrey R. Kling, "Incarceration Length, Employment, and Earnings," *Am. Econ. Rev.* (2006).

72: **Recidivism rates:** Danton Berube and Donald P. Green, "Do Harsher Sentences Reduce Recidivism? Evidence from a Natural Experiment" (2007), working paper.

73: **The Poverty Action Lab:** J. R. Minkel, "Trials for the Poor: Rise of Randomized Trials to Study Antipoverty Programs," *Sci. Am.* (Nov. 21, 2005), http://www.sciam.com/article.cfm?articleID=000ECBA5-A101-137B-A10183414B7F0000.

74: **Female chiefs:** Raghabendra Chattopadhyay and Esther Duflo, "Women's Leadership and Policy Decisions: Evidence from a Nationwide Randomized Experiment in India," Inst. Econ. Dev. Paper No. dp-114 (2001), http://www.bu.edu/econ/ied/dp/papers/chick3.pdf.

74–75: **Testing teacher absenteeism:** Esther Duflo and Rema Hanna, "Monitoring Works: Getting Teachers to Come to School," Nat'l Bureau of Econ. Research Working Paper No. 11880 (2005); Swaminathan S. A. Aiyar, "Camera Schools: The Way to Go," *Times of India*, Mar. 11, 2006, http://timesofindia.india times.com/articleshow/1446353.cms.

75: **Kenyan de-worming:** Michael Kremer and Edward Miguel, "Worms: Education and Health Externalities in Kenya," Poverty Action Lab Paper No. 6 (2001), http://www.povertyactionlab.org/papers/kremer_miguel.pdf.

75: **Auditing in Indonesia:** Benjamin A. Olken, "Monitoring Corruption: Evidence from a Field Experiment in Indonesia," Nat'l Bureau of Econ. Research Working Paper No. 11753 (2005); "Digging for Dirt," *Economist*, Mar. 16, 2006.

75–79: **Progresa's poverty program:** Paul Gertler, "Do Conditional Cash Transfers Improve Child Health?: Evidence from PROGRESA's Control Randomized Experiment," 94 *Am. Econ. Rev.* 336 (2004).

75: **Paying mothers vs. fathers:** Gertler stressed that targeting mothers may not be appropriate in other cultures. "In fact, in Yemen right now we are doing a conditional cash transfer where we are randomizing giving the money to the mother vs. giving the money to the father to see whether it makes a difference." See Paul Gertler, "Do Conditional Cash Transfers Improve Child Health?: Evidence from PROGRESA's Control Randomized Experiment," 94 *Am. Econ. Rev.* 336 (2004).

CHAPTER 4

Page 81: **The beginning of EBM:** Gordon Guyatt et al., "Evidence-Based Medicine: A New Approach to Teaching the Practice of Medicine," 268 *JAMA* 2420 (1992); "Glossary of Terms in Evidence-Based Medicine," Oxford Center for Evidence-Based Medicine, http://www.cebm.net/glossary.asp; Gordon Guyatt et al., "Users' Guide to the Medical Literature: Evidence-Based Medicine: Principles for Applying the Users' Guides to Patient Care," 284 *JAMA* 1290 (2000).

82: **Ignaz Semmelweis takes the next step in medical Super Crunching:** "The Cover: Ignaz Philipp Semmelweis (1818–65)," 7 *Emerging Infectious Diseases,* cover page (Mar.–Apr. 2001), http://www.cdc.gov/ncidod/eid/vol7no2/cover.htm.

82: **Pasteur also early supporter of hand-washing:** Louis Pasteur was, however, convinced by Semmelweis's results. In a speech before the Academy of Medicine, Pasteur opined: "If I had the honor of being a surgeon ... not only would I use none but perfectly clean instruments, but I would clean my hands with the greatest care. ..." Theodore L. Brown, *Science and Authority* (book manuscript, 2006).

83: **Don Berwick, the modern-day Semmelweis:** Tom Peters, "Wish I Hadn't Read This Today," *Dispatches from the World of Work,* Dec. 7, 2005, http://tompeters.com/entries.php?note=008407.php; Inst. of Med., *To Err Is Human: Building a Safer Health System* (1999); Donald Goldmann, "System Failure Versus Personal Accountability—The Case for Clean Hands," 355 *N. Engl. J. Med.* 121 (Jul. 13, 2006); Neil Swidey, "The Revolutionary," *Boston Globe,* Jan. 4, 2004; Donald M. Berwick et al., "The 100,000 Lives Campaign: Setting a Goal and a Deadline for Improving Health Care Quality," 295 *JAMA* 324 (2006); "To the Editor," *N. Engl. J. Med.* (Nov. 10, 2005).

85: **Reducing risk of central-line catheter infections:** S. M. Berenholtz, P. J. Pronovost, P. A. Lipsett, et al. "Eliminating Catheter-Related Bloodstream Infections in the Intensive Care Unit," *Crit. Care Med.* 32 (2004), pp. 2014–2020.

87–88: **Medical myths and practices that won't die:** Gina Kolata, "Annual Physical Checkup May Be an Empty Ritual," *N.Y. Times,* Aug. 12, 2003, p. F1; Douglas S. Paauw, "Did We Learn Evidence-Based Medicine in Medical School? Some Common Medical Mythology," 12 *J. Am. Bd. Family Practice* 143 (1999); Robert J. Flaherty, "Medical Mythology," http://www.montana.edu/wwwebm/myths/home.htm.

88: **Can alternative medicine be evidence-based?:** The resistance to statistical evidence is even more pronounced with regard to "alternative medicine," a vast field that encompasses everything from herbal supplements to energy therapy to yoga. More than a third of American adults use some form of alternative medicine annually; they spend more than $40 billion a year on alternative treatments. Patricia M. Barnes, et al., "Complementary and Alternative Medicine Use Among Adults," CDC Advance Data Report No. 343 (May 27, 2004), available at http://www.cdc.gov/nchs/data/ad/ad343.pdf.

Many adherents and advocates of alternative medicine reject not only Western treatments but the Westernized notion of statistical testing. They sometimes claim that their practices are too "individual" or "holistic" to study scientifically and instead rely on anecdotes and case studies without adequate controls or control groups for comparison. I'm agnostic about whether alternative medicine is effective. But it verges on idiocy to claim that the effectiveness cannot be tested. If it really is important, as alternative medicine advocates claim, to take into account a larger set of information about the patient ("what kind of person has the disease"), then providers who do so should produce better results. There's no reason why Super Crunching can't be used to test whether alternative medicine works.

To support EBM doesn't mean that you're hostile to alternative or innovative—even seemingly wacky—therapies. When it comes to the back-end inquiry of finding out which treatments are effective, there is no East and West. I throw my lot with two past editors-in-chief of the *New England Journal of Medicine,* Marcia Angell and Jerome Kassirer:

> It is time for the scientific community to stop giving alternative medicine a free ride. There cannot be two kinds of medicine—conventional and alternative. There is only medicine that has been adequately tested and medicine that has not, medicine that works and medicine that may or may not work.

M. Angell and J. P. Kassirer, "Alternative Medicine: The Risks of Untested and Unregulated Remedies," 339 *N. Engl. J. Med.* 839 (1998). See also Kirstin Borgerson, "Evidence-Based Alternative Medicine?" 48 *Perspectives in Bio. and Med.* 502 (2005); W. B. Jonas, "Alternative Medicine: Learning from the Past,

Examining the Present, Advancing to the Future," 280 *JAMA* 1616 (1998); P. B. Fontanarosa and G. D. Lundberg, "Alternative Medicine Meets Science," 280 *JAMA* 1618 (1998).

88: **The Aristotelian approach:** Kevin Patterson, "What Doctors Don't Know (Almost Everything)," *N.Y. Times Magazine*, May 5, 2002; David Leonhardt, "Economix: What Money Doesn't Buy in Health Care," *N.Y. Times*, Dec. 13, 2006.

89: **Contemporary uses (or lack thereof) of EBM:** Brandi White, "Making Evidence-Based Medicine Doable in Everyday Practice," *Family Practice Mgt.* (Feb. 2004), http://www.aafp.org/fpm/20040200/51maki.html; Jacqueline B. Persons and Aaron T. Beck, "Should Clinicians Rely on Expert Opinion or Empirical Findings?" 4 *Am. J. Mgd Care* 1051 (1998), http://www.ajmc.com/files/article files/AJMC1998JulPersons1051_1054.pdf; D. G. Covell et al., "Information Needs in Office Practice: Are They Being Met?" 103 *Ann. Internal Med.* 596 (1985); John W. Ely et al., "Analysis of Questions Asked by Family Doctors Regarding Patient Care," 319 *BMJ* 358 (1999); "The Computer Will See You Now," *Economist*, Dec. 8, 2005, http://www.microsoft.com/business/peopleready/news/economist/computer.mspx.

90: **Coffee and Heart Disease:** A. Z. LaCroix et al., "Coffee Consumption and the Incidence of Coronary Heart Disease," 315 *N. Engl. J. Med.* 977 (1986); Int'l Food Info. Council Foundation, "Caffeine and Health: Clarifying the Controversies" (1993), www.familyhaven.com/health/ir-caffh.html.

91: **17 years:** Institute of Medicine, Committee on Quality of Health Care in America, *Crossing the Quality Chasm: A New Health System for the 21st Century*, National Academy Press, Washington, D.C. (2001).

91: **One funeral at a time:** Various formulations of this quotation have been attributed to Max Planck, Albert Einstein, and Paul Samuelson.

92–93: **Few doctors rely on research to treat individual patients:** Brandi White, "Making Evidence-Based Medicine Doable in Everyday Practice," *Family Practice Mgt.* (Feb. 2004), http://www.aafp.org/fpm/20040200/51maki.html; D. M. Windish and M. Diener-West, "A Clinician-Educator's Roadmap to Choosing and Interpreting Statistical Tests," 21 *J. Gen. Intern. Med.* 656 (2006); Kevin Patterson, "What Doctors Don't Know (Almost Everything)," *N.Y. Times Magazine*, May 5, 2002.

92–93: **Where is the library in the doctor's office?:** A law library is physically and intellectually at the heart of many law practices. A client comes to a lawyer for advice, and the lawyer hits the books to find out what the law is for that client's particular problem. Until evidence-based medicine came along, most doctors didn't routinely research the problems of individual patients. The physician might try to keep up in the field generally. He or she might subscribe to the *New England Journal of Medicine* and *JAMA*. Yet rare would be the case where a physician would pick up a book or journal article to do patient-specific research. In sharp contrast to a law office, most doctors' offices never had a library. If a physician didn't know the answer, she might consult with a specialist, but neither the physician nor the specialist was very likely to pick up a book and read.

93: **Information must be retrievable to be useful:** Gordon Guyatt et al., "Evidence-Based Medicine: A New Approach to Teaching the Practice of Medicine," 268 *JAMA* 2420 (1992); "Center for Health Evidence, Evidence-Based Medicine: A New Approach to Teaching the Practice of Medicine" (2001), http://www.cche.net/usersguides/ebm.asp; Lisa Sanders, "Medicine's Progress, One Setback at a Time," *N.Y. Times*, Mar. 16, 2003, p. 29; InfoPOEMs, https://www.infopoems.com.

96: **Data-based diagnostic decisions:** "To the Editor," *N. Engl. J. Med.*, Nov. 10, 2005.

97: **Diagnostic-decision support software:** The major softwares in this field include Isabel, QMR, Iliad, Dxplain, DiagnosisPro, and PKC.

97–99: **Super Crunching away diagnostic errors:** Jeanette Borzo, "Software for Symptoms," *Wall St. J. (Office Technology)*, May 23, 2005, p. R10; Jason Maude biography, The Beacon Charitable Trust, http://www.beaconfellowship.org.uk/biography2003_11.asp; David Leonhardt, "Why Doctors So Often Get It Wrong," *N.Y. Times*, Feb. 22, 2006.

99: **Isabel helps doctors consider major diagnosis:** Stephen M. Borowitz, et al., "Impact of a Web-based Diagnosis Reminder System on Errors of Diagnosis," presented at AMIA 2006: Biomedical and Health Informatics (Nov. 11, 2006).

CHAPTER 5

Pages 104–108: **Super Crunchers take on experts to predict Supreme Court decisions:** Andrew D. Martin et al., "Competing Approaches to Predicting Supreme Court Decision Making," 2 *Persp. on Pol.*, 763 (2004); Theodore W. Ruger et al., "The Supreme Court Forecasting Project: Legal and Political Science Approaches to Predicting Supreme Court Decisionmaking," 104 *Colum. L. Rev.* 1150 (2004).

105: **Holmes on legal positivism:** Oliver W. Holmes, *The Common Law* 1 (1881) ("The prophesies of what the courts will do in fact, and nothing more pretentious, are what I mean by the law."). See also Oliver Wendell Holmes, Jr., "The Path of the Law," 10 *Harv. L. Rev.* 457, 461 (1897) ("The object of our study, then, is prediction, the prediction of the incidence of the public force through the instrumentality of the courts.").

105: **Langdell on law as a science:** Christopher C. Langdell, "Harvard Celebration Speeches," 3 *L.Q. Rev.* 123, 124 (1887).

108–111: **Meehl's little book (and other works):** Paul E. Meehl, *Clinical Versus Statistical Prediction: A Theoretical Analysis and a Review of the Evidence* (1954). See also William M. Grove, "Clinical Versus Statistical Prediction: The Contribution of Paul E. Meehl," 61 *J. Clinical Psychol.* 1233 (2005), http://www.psych. umn.edu/faculty/grove/112clinicalversusstatisticalprediction.pdf; Michael P. Wittman, "A Scale for Measuring Prognosis in Schizophrenic Patients," 4 *Elgin Papers* 20 (1941); Drew Western and Joel Weinberger, "In Praise of Clinical Judgment: Meehl's Forgotten Legacy," 61 *J. Clinical Psychol.* 1257, 1259 (2005), http://www.psychsystems.net/lab/2005_w_weinberger_meeh_JCP.pdf; Paul E. Meehl, in 8 *A History of Psychology in Autobiography* 337, 354 (G. Lindzey, ed., 1989).

110–111: **Snijders vs. the buying experts:** Chris Snijders et al., "Electronic Decision Support for Procurement Management: Evidence on Whether Computers Can Make Better Procurement Decisions," 9 *J. Purchasing and Supply Mgmt* 191 (2003); Douglas Heingartner, "Maybe We Should Leave That Up to the Computer," *N.Y. Times,* Jul. 18, 2006.

111–112: **Man vs. machine meta analysis:** William M. Grove and Paul E. Meehl, "Comparative Efficiency of Informal (Subjective, Impressionistic) and Formal (Mechanical, Algorithmic) Prediction Procedures: The Clinical–Statistical Controversy," 2 *Psychol. Pub. Pol'y and L.* 293, 298 (1996); William M. Grove, "Clinical Versus Statistical Prediction: The Contribution of Paul E. Meehl," 61 *J. Clinical Psychol.* 1233 (2005), http://www.psych.umn.edu/faculty/grove/112 clinicalversusstatisticalprediction.pdf.

112: **Human bias:** D. Kahneman et al., *Judgment Under Uncertainty: Heuristics and Biases* (1982); R. M. Dawes and M. Mulford, "The False Consensus Effect and Overconfidence: Flaws in Judgment, or Flaws in How We Study Judgment?" 65 *Organizational Behavior and Human Decision Processes* 201 (1996).

112: **A pool is more dangerous than a gun:** Steven Levitt, editorial, "Pools More Dangerous Than Guns," *Chi. Sun-Times,* Jul. 28, 2001, p. 15.

114: **People guess poorly:** J. Edward Russo and Paul J. H. Schoemaker, *Decision Traps: Ten Barriers to Brilliant Decision-Making and How to Overcome Them* (1990). See also Scott Plous, *The Psychology of Judgment and Decision Making* (2003); John Ruscio, "The Perils of Post-Hockery: Interpretations of Alleged Phenomena After the Fact," *Skeptical Inquirer,* Nov.–Dec. 1998.

114: **Costs of the Iraq War:** Interview with Vice President Cheney, CNN broadcast, Jun. 20, 2005;. interview with Glenn Hubbard, CNBC broadcast, October 4, 2002; Reuters, "U.S. Officials Play Down Iraq Reconstruction Needs," *Entous*, Apr. 11, 2003; Hearing on a Supplemental War Regulation Before the H. Comm. on Appropriations, 108th Cong. (Mar. 27, 2003) (statement of Deputy Defense Sec'y Paul Wolfowitz); Rep. Jan Schakowsky, "Past Comments About How Much Iraq Would Cost," www.house.gov/schakowsky/iraqquotes_web.htm.

115: **Human judges:** Richard Nisbett and Lee Ross, *Human Inference: Strategies and Shortcomings of Social Judgment* (1980).

115: **Super Crunching without emotions:** Douglas Heingartner, "Maybe We Should Leave That Up to the Computer," *N.Y. Times*, Jul. 18, 2006.

116–117: **Can Super Crunchers and experts coexist?:** S. Schwartz et al., "Clinical Expert Systems Versus Linear-Models: Do We Really Have to Choose," 34 *Behavioral Sci.* 305 (1989).

118: **Humans serving machines:** Douglas Heingartner, "Maybe We Should Leave That Up to the Computer," *N.Y. Times*, Jul. 18, 2006.

118: **Parole predictions:** For an early study assessing the likely outcomes for parolees, see Earnest W. Burgess, "Factors Determining Success or Failure on Parole," in *The Workings of the Indeterminate Sentence Law and the Parole System in Illinois* (A. A. Bruce, ed., 1928), pp. 205–249. For biographical information and account of Burgess's place in sociology, use of new methods of measurement, see Howard W. Odum, *American Sociology: The Story of Sociology in the United States Through 1950* (1951), pp.168–171, http://www2.asanet.org/governance/burgess.html.

119: **Clouston and the SVPA:** Virginia General Statutes, § 37.1–70.4 (C); Frank Green, "Where Is This Man?: Should This Child Molester and Cop Killer Have Been Released?" *Richmond Times-Dispatch*, Apr. 16, 2006.

119: **Supremes uphold SVPA:** *Kansas v. Hendricks*, 521 U.S. 346 (1997).

119: **Monahan on the first Super Crunching statutory trigger:** Bernard E. Harcourt, *Against Prediction: Profiling, Policing and Punishment in an Actuarial Age* (2007); John Monahan, *Forecasting Harm: The Law and Science of Risk Assessment among Prisoners, Predators, and Patients*, ExpressO Preprint Series (2004), http://law.bepress.com/expresso/eps/410.

121: **Conditioning commitment on legal, but statistically predictive, behavior:** Eugene Volokh on his blog, The Volokh Conspiracy, has questioned whether the sex discrimination would withstand constitutional scrutiny. Eugene Volokh, "Sex Crime and Sex," The Volokh Conspiracy, Jul. 14, 2005, http://volokh.com/posts/1121383012.shtml. I do have concerns about whether the RRASOR system is in fact statistically valid. The original dataset upon which it was estimated was "artificially reduced" to 1,000 observations to intentionally reduce the statistical significance of the explanatory variables. R. Karl Hanson, *The Development of a Brief Actuarial Scale for Sexual Offense Recidivism* (1997). Instead of constructing the crude point system, a more tailored approach would estimate recidivism probabilities based directly on the coefficients of the regression equation. The constructors of RRASOR seemed to be operating in a pre-computer environment where assessments had to be easily calculated by hand.

121: **Introduction of the broken-leg case:** Paul E. Meehl, *Clinical Versus Statistical Prediction: A Theoretical Analysis and a Review of the Evidence* (1954) (reissued University of Minnesota 1996).

122: **Tom Wolfe:** Tom Wolfe, *The Right Stuff* (1979).

122–23: **Mistaken discretionary releases:** Frank Green, "Where Is This Man?: Should This Child Molester and Cop Killer Have Been Released?" *Richmond Times-Dispatch*, Apr. 16, 2006.

123–124: **Assessing human overrides:** James M. Byrne and April Pattavina, "Assessing the Role of Clinical and Actuarial Risk Assessment in an Evidence-Based Community Corrections System: Issues to Consider," *Fed. Probation* (Sep. 2006). See also Laurence L. Motiuk et al., "Federal Offender Population Movement: A Study of Minimum-security Placements," Correctional Service of Canada, (Mar. 2001); Patricia M. Harris, "What Community Supervision Officers Need to Know About Actuarial Risk Assessment and Clinical Judgment," *Fed. Probation* (Sep. 2006).

124–125: **The work that's left for humans:** Drew Western and Joel Weinberger, "In Praise of Clinical Judgment: Meehl's Forgotten Legacy," 61 *J. Clinical Psychol.* 1257, 1259 (2005); Paul E. Meehl, "What Can the Clinician Do Well?" in *Problems in Human Assessment* 594 (D. N. Jackson and S. Messick, eds., 1967); Paul E. Meehl, "Causes and Effects of My Disturbing Little Book," 50 *J. Personality Assessment* 370 (1986).

124: **Fink's circumcision hypothesis:** A. J. Fink, Letter, "A Possible Explanation for Heterosexual Male Infection with AIDS," 315 *N. Engl. J. Med.* 1167 (1986).

125–126: **HIV-transmission hypotheses:** Cameron D. William et al., "Female to Male Transmission of Human Immunodeficiency Virus Type 1: Risk Factors for Seroconversion in Men," 2 *Lancet* 403 (1989). See also J. Simonsen et al., "Human Immunodeficiency Virus Infection in Men with Sexually Transmitted Diseases," 319 *N. Engl. J. Med.* 274 (1988); M. Fischl et al., "Seroprevalence and Risks of HIV Infection in Spouses of Persons Infected with HIV," Book 1, 4th Int'l Conf. on AIDS 274, Stockholm, Sweden (Jun. 12–16, 1988).

125: **Continuing empiricism on the circumcision-AIDS connection:** D. T. Halperin et al., "Male Circumcision and HIV Infection: 10 Years and Counting,"

354 *Lancet* 1813 (1999); Donald G. McNeil Jr., "Circumcision's Anti-AIDS Effect Found Greater Than First Thought," *N.Y. Times*, Feb. 23, 2007.

126: **Weight loss investments:** I personally have signed a contract putting thousands of my dollars at risk if I don't take off twenty pounds this year and keep it off. Dean Karlan and I are trying to put together a randomized study of weight-loss bonds. Send me an email at ian.ayres@yale.edu if you're interested in participating.

127–128: **Hammond on clinical resistance:** Kenneth R. Hammond, *Human Judgment and Social Policy* 137–38 (1996).

CHAPTER 6

Page 130: **Google Books creates a virtual library:** Jeffrey Toobin, "Google's Moon Shot," *New Yorker*, Feb. 5, 2007.

130–131: **Discriminatory negotiation practices:** These results came from my initial study: Ian Ayres, "Fair Driving: Gender and Race Discrimination in Retail Car Negotiations," 104 *Harv. L. Rev.* 817 (1991). An analysis of five more recent studies can be found in my book *Pervasive Prejudice?: Non-Traditional Evidence of Race and Gender Discrimination* (2001).

132–133: **Discriminatory auto-lending mark-ups:** Ian Ayres, "Market Power and Inequality: A Competitive Conduct Standard for Assessing When Disparate Impacts Are Justified," *Cal. L. Rev.* (2007).

134–135: **ChoicePoint sells data:** Gary Rivlin, "Keeping Your Enemies Close," *N.Y. Times*, Nov. 12, 2006; ChoicePoint 2005 Annual Report, http://library.corporate-ir.net/library/95/952/95293/items/189639/2005annual.pdf.

134: **Biggest company you never heard of:** "Persuaders," PBS *Frontline*, http://www.pbs.org/wgbh/pages/frontline/shows/persuaders/etc/script.html; see also Richard Behar, "Never Heard of Acxiom? Chances Are It's Heard of You," *Fortune*, Feb. 23, 2004.

134: **850 terabytes of storage:** Rick Whiting, "Tower of Power," *InformationWeek*, Feb. 11, 2002.

135: **Data silos:** Kim Nash, "Merging Data Silos," *Computerworld*, April 15, 2002; Gary Rivlin, "Keeping Your Enemies Close," *N.Y. Times*, Nov. 12, 2006; Eric K. Neumann, "Freeing Data, Keeping Structure," *Bio-IT World*, Jun. 14, 2006.

136: **Data "mashups":** Rachel Rosmarin, "Maps. Mash-ups. Money." Forbes.com, Jun. 16, 2006, http://www.forbes.com/technology/2006/06/14/google-yahoo-microsoft_cx_rr_0615maps.html.

136: **Car theft database:** This data is available from the Federal Bureau of

Investigation's National Crime Information Center (NCIC), http://www.fas.org/irp/agency/doj/fbi/is/ncic.htm.

137–138: **Erroneous felon disenfranchisement:** The United States Civil Rights Commission, *The 2000 Presidential Elections* (www.usccr.gov/pubs/vote2000/report/ch5.htm).

139: **Kryder's Law:** "Kryder's Law" is the title of an article by Chip Walter in the Aug. 2005 issue of *Scientific American.*

140: **Computer storage capacity costs decrease over time:** "Historical Notes about the Cost of Hard Drive Storage Space," www.littletechshoppe.com/ns1625/winchest.html; Jim Handy, "Flash Memory vs. Hard Disk Drives— Which Will Win?" Jun. 6, 2005, http://www.storagesearch.com/semico-art1.html.

140: **Twelve terabytes for Yahoo:** Kevin J. Delaney, "Lab Test: Hoping to Overtake Its Rivals, Yahoo Stocks Up on Academics," *Wall St. J.,* Aug. 25, 2006, p. A1.

141–144: **Background on neural networks:** P. L. Brockett et al., "A Neural Network Method for Obtaining an Early Warning of Insurer Insolvency," 61 *J. Risk and Insurance* 402 (1994); Jack V. Tu, "Advantages and Disadvantages of Using Artificial Neural Networks versus Logistic Regression for Predicting Medical Outcomes," 50 *J. Clin. Epidemiol.* 1309 (1996); W. G. Baxt, "Analysis of the Clinical Variables Driving Decision in an Artificial Neural Network Trained to Identify the Presence of Myocardial Infarction," 21 *Ann. Emerg. Med.* 1439 (1992); K. A. Spackman, "Combining Logistic Regression and Neural Networks to Create Predictive Models," in *Proceedings of the Sixteenth Annual Symposium on Computer Applications in Medical Care* (1992), pp. 456–59; J. L. Griffith et al., "Statistical Regression Techniques for the Construction, Interpretation and Testing of Computer Neural Networks," 12 *Med. Decis. Making* 343 (1992).

142–143: **Neural networks and greyhound racing:** Hsinchun Chen et al., "Expert Prediction, Symbolic Learning, and Neural Networks: An Experiment on Greyhound Racing," 9 *IEEE Expert* 21 (Dec. 1994).

145: **Epagogix revealed:** Malcolm Gladwell, "The Formula," *New Yorker,* Oct. 16, 2006.

151–152: **Lulu scores book titles:** Misty Harris, "Anyone Who Says You Can't Judge a Book by Its Cover Isn't Trying Hard Enough," *Windsor Star,* Dec. 17, 2005, http://www.canada.com/windsorstar/features/onlineextras/story.html?id=35711d6b-f13e-47d9-aa23-7eaa12bc8846.

152–153: **What makes a law review article likely to be cited?:** Ian Ayres and Fredrick E. Vars, "Determinants of Citations to Articles in Elite Law Reviews," 29 *J. Legal Stud.* 427, 433–34 (2000).

153: **Just stick the right formula in:** Indigo Girls, "Least Complicated," on *Swamp Ophelia* (1994).

154: **Robert Frost on walls:** Robert Frost famously wrote, "Something there is that doesn't love a wall/That wants it down." Robert Frost, "Mending Wall" (1915).

154: **Library of Alexandria several times over on the Internet:** Kevin Kelly, "Scan this Book!" *N.Y. Times Magazine,* May 14, 2006, p. 43.

155: **Ubiquitous surveillance:** A. Michael Froomkin, "The Death of Privacy," 52 *Stan. L. Rev.* 1461 (2000).

155: **Nanotechnological advances:** George Elvin, "The Coming Age of Nanosensors," *Nanotech\Buzz,* http://www.nanotechbuzz.com/50226711/the_coming_age_of_nanosensors.php.

155: **Smart dust:** Gregor Wolbring, "The Choice Is Yours: Smart Dust," *Innovation Watch,* Dec. 15, 2006, http://www.innovationwatch.com/choiceisyours/choiceisyours.2006.12.15.htm.

CHAPTER 7

Pages 156–158: **Ms. Daniel's lesson:** Siegfried Engelmann and Elaine C. Bruner, "The Pet Goat," in *Reading Mastery II: Storybook 1* (1997). See also Daniel Radosh, "The Pet Goat Approach," *New Yorker,* Jul. 26, 2004.

158: **The goat story on the silver screen:** *Fahrenheit 9/11* (Sony Pictures 2004); Daniel Radosh, "The Pet Goat Approach," *New Yorker,* Jul. 26, 2004.

158–169: **The controversy over Direct Instruction:** W. C. Becker, "Direct Instruction: A Twenty Year Review," in *Designs for Excellence in Education: The Legacy of B. F. Skinner* (R. P. West and L. A. Hamerlynck, eds., 1992), pp. 71–112; G. L. Adams and S. Engelmann, *Research on Direct Instruction: 25 Years beyond DISTAR* (1996); Am. Fed. Teachers, *Direct Instruction* (1998); American Institutes for Research, Comprehensive School Reform Quality Center, *CSRQ Center Report on Elementary School Comprehensive School Reform Models* (2006), http://www.csrq.org/documents/CSRQCenterCombinedReport_Web11-03-06.pdf; Jean Piaget, *Adaptation and Intelligence: Organic Selection and Phenocopy* (1980); Linda B. Stebbins et al., *Education as Experimentation: A Planned Variation Model Volume IV-A An Evaluation of Project Follow Through* (1977); Sanjay Baht, "A New Way of Judging How Well Schools Are Doing," *Seattle Times,* Aug. 2, 2005; Ted Hershberg et al., "The Revelations of Value-Added," *School Administrator,* Dec. 2004; Siegfried Engelmann, *War Against the Schools' Academic Child Abuse* (1992); David Glenn, "No Classroom Left Unstudied," *The Chronicle of Higher Education,*

May 28, 2004; Richard Nadler, "Failing Grade," *Nat'l Rev.*, Jun. 1, 1998; Am. Fed. Teachers, *Building on the Best, Learning from What Works: Six Promising Schoolwide Reform Programs* (1998); B. Gunn et al., "The Efficacy of Supplemental Instruction in Decoding Skills for Hispanic and Non-Hispanic Students in Early Elementary School," 34 *J. Special Ed.* 90 (2000); B. Gunn et al., "Supplemental Instruction in Decoding Skills for Hispanic and Non-Hispanic Students in Early Elementary School: A Follow-Up," 36 *J. Special Ed.* 69 (2002).

159: **Some samples of Engelmann's work:** G. L. Adams and S. Engelmann, *Research on Direct Instruction: 25 Years beyond DISTAR* (1996); Siegfried Engelmann, *War Against the Schools' Academic Child Abuse* (1992). See also Daniel Radosh, "The Pet Goat Approach," *New Yorker*, Jul. 26, 2004.

159: **Piaget's child-centered approach:** Jean Piaget, *Adaptation and Intelligence: Organic Selection and Phenocopy* (1980).

160–161: **Project Follow Through:** Bonnie Grossen (ed.), "Overview: The Story Behind Project Follow Through," *Effective School Practices*, 15:1, Winter 1995–96, http://darkwing.uoregon.edu/~adiep/ft/grossen.htm.

161: **DI wins hands down:** Richard Nadler, "Failing Grade," *Nat'l Rev.*, Jun. 1, 1998, http://www.nationalreview.com/01jun98/nadler060198.html.

161: **Recent studies support DI:** Am. Fed. Teachers, *Building on the Best, Learning from What Works: Six Promising Schoolwide Reform Programs* (1998); B. Gunn et al., "The Efficacy of Supplemental Instruction in Decoding Skills for Hispanic and Non-Hispanic Students in Early Elementary School," 34 *J. Special Ed.* 90 (2000); B. Gunn et al., "Supplemental Instruction in Decoding Skills for Hispanic and Non-Hispanic Students in Early Elementary School: A Follow-Up," 36 *J. Special Ed.* 69 (2002); Angela M. Przychodzin, "The Research Base for Direct Instruction in Mathematics," SRA/McGraw-Hill, https://www.sraonline.com/download/DI/Research/Mathematics/research_base_for%20di_math.pdf.

163: **The Arundel DI experiment:** "A Direct Challenge," *Ed. Week*, Mar. 17, 1999, http://www.zigsite.com/DirectChallenge.htm; Martin A. Kozloff et al., *Direct Instruction in Education*, Jan. 1999, http://people.uncw.edu/kozloffm/diarticle.html; Nat'l Inst. Direct Instruction, http://www.nifdi.org; Daniel Radosh, "The Pet Goat Approach," *New Yorker*, Jul. 26, 2004.

164: **The Michigan study:** L. Schweinhart et al., "Child-Initiated Activities in Early Childhood Programs May Help Prevent Delinquency," 1 *Early Child. Res. Q.* 303–312 (1986). A larger proportion of the students who were taught using DI were male and were more likely to stay in state. These students might have been arrested more often in Michigan not because of DI but because men are more likely to commit crime and because people who were arrested out of state were not considered in the analysis. See also Paulette E. Mills et al., "Early Exposure to

Direct Instruction and Subsequent Juvenile Delinquency: A Prospective Examination," 69 *Exceptional Child*, 85–96 (2002) (finding no impact of Direct Instruction on subsequent juvenile delinquency).

165: **The Bush administration approach:** Daniel Radosh, "The Pet Goat Approach," *New Yorker*, Jul. 26, 2004; U.S. Dep't of Ed., What Works Clearinghouse, http://www.whatworks.ed.gov/; Southwest Ed. Dev. Lab., "What Does a Balanced Approach Mean?" http://www.sedl.org/reading/topics/balanced.html.

165: **"Scientifically based" programs:** The term was first found in the 1990s, promoting funding for "scientifically based" education research.

165: **The president's support of DI:** In "The Pet Goat Approach," Daniel Radosh speculated about a less pristine motive for our president's support of the Direct Instruction curriculum, which is published by McGraw-Hill: "[I]t's easy to imagine one of [Michael] Moore's hallmark montages, spinning circumstantial evidence into a conspirational web: a sepia-toned photograph from the thirties of, say, Prescott Bush and James McGraw, Jr., palling around on Florida's Jupiter Island; a film clip from the eighties of Harold McGraw, Jr., joining the advisory panel of Barbara Bush's literacy foundation; Harold McGraw III posing with President George W. Bush as part of his transition team; and, to tie it all together, former McGraw-Hill executive vice-president John Negroponte being sworn in as the new Ambassador to Iraq." Daniel Radosh, "The Pet Goat Approach," *New Yorker*, Jul. 26, 2004.

167: **Loan decisions by the numbers:** Peter Chalos, "The Superior Performance of Loan Review Committee," 68 *J. Comm. Bank Lending* 60 (1985).

167: **Disabling discretion:** The value of disabling discretion isn't just for the other guy. When we make snap judgments, we sometimes lose control of unconscious influences that can seep into our decision-making process. See generally Malcolm Gladwell, *Blink: The Power of Thinking Without Thinking* (2005). A few years ago, I collected information on how people tipped cab drivers in New Haven. Ian Ayres et al., "To Insure Prejudice: Racial Disparities in Taxicab Tipping," 114 *Yale L. J.* 1613 (2005). A disheartening finding was that passengers tipped minority drivers about a third less than white drivers providing the same-quality service. At first, I was pretty sure that I didn't tip minorities less because with very rare exceptions I always tip 20 percent. When I took a closer look at the data, however, I found that a lot of the passenger discrimination came from a seemingly innocuous source. Cab passengers like to round the total amount that they pay to a round number. We've all had the experience. Just before your cab arrives the fare clicks over to $7. Think fast. Do you leave $8 or $9? We suspect that passengers are often called upon to make a quick decision about whether to round up or down. Even people who think that they are hard-wired 20 percent tippers may find that unconscious factors influence this rounding decision. In my study, passengers were much more likely to round up with white drivers and more likely to round down with minority drivers. Suddenly, I'm not so confident that "I *certainly* don't discriminate." I'm more of a discretionary tipper than I had ever imagined.

168: **Decline of physician status:** Kevin Patterson, "What Doctors Don't Know (Almost Everything)," *N.Y. Times Magazine,* May 5, 2002, p. 74.

168: **Along Came Polly:** A trailer with quotation viewable at http://www. apple.com/trailers/universal/along_came_polly/medium.html.

169: **Lies, damn lies:** This phrase was popularized in the U.S. by Mark Twain, but is attributed to English statesman Benjamin Disraeli.

173: **Pain points:** Harrah's prediction of customer pain points was discussed in Chapter 1. See also Christopher Caggiano, "Show Me the Loyalty," *CMO Magazine,* Oct. 2004.

174–175: **Virtual redlining in insurance?:** Disclosure: I was a paid economics expert for minority plaintiffs in this case (as well as in other cases bringing similar claims). *Owens v. Nationwide Mutual Ins. Co.,* No. Civ. 3:03-CV-1184-H, 2005 WL 1837959 (N.D. Tex. Aug. 2, 2005) (holding that even if its use of credit scores had a disparate impact on minorities, Nationwide had legitimate, race-neutral business reasons for using credit scores). See also *Powell v. American General Finance, Inc.,* 310 F. Supp. 2d 481 (N.D.N.Y. 2004).

175: **Justice O'Connor urging "race-neutral means to increase minority participation":** *Adarand Constructors, Inc. v. Pena,* 515 U.S. 200, 238 (1995); *City of Richmond v. J. A. Croson Co.,* 488 U.S. 469, 507 (1989).

176: **Veterans' virtual records disappeared:** David Stout and Tom Zeller, Jr., "Vast Data Cache About Veterans Has Been Stolen," *N.Y. Times,* May 23, 2006, p. A1.

176: **Fidelity laptop stolen:** Jennifer Levitz and John Hechinger, "Laptops Prove Weakest Link in Data Security," *Wall St. J.,* Mar. 24, 2006, p. B1.

176: **AOL user information released:** Saul Hansell, "AOL Removes Search Data on Vast Group of Web Users," *N.Y. Times,* Aug. 8, 2006, p. C4.

177: **Frost's definition of home:** Robert Frost, "The Death of the Hired Man," in *North of Boston* (1914).

177: **The end of anonymity:** Jed Rubenfeld, "Privacy's End" (working paper October 2006).

177: **New facial recognition innovations:** Anick Jesdanun, "Facial-ID Tech and Humans Seen as Key to Better Photo Search, But Privacy Concerns Raised," Associated Press, Dec. 28, 2006.

179: Gattaca's **vision of genetic predestination:** The Science Show, http://www.abc.net.au/rn/scienceshow/stories/2001/262366.htm.

179: **Google's mission statement:** Google Corporate Information: Company Overview, http://www.google.com/corporate/.

180: **Americans don't protect their privacy:** Bob Sullivan, "Privacy Under Attack, But Does Anybody Care?" MSNBC, Oct. 17, 2006, http://www.msnbc.msn.com/id/15221095/.

180: **Mary Rosh criticizes Ayres and Donohue:** Tim Lambert, Mary Rosh's Blog, http://timlambert.org/2003/01/maryrosh/.

180–181: **Lott's claims on guns and crime:** John Lott, *More Guns, Less Crime* (2000).

181: **Ayres and Donohue response article to Lott:** Ian Ayres and John J. Donohue III, "Shooting Down the 'More Guns, Less Crime' Hypothesis," 55 *Stanford L. Rev.* 1193 (2003); Ian Ayres and John J. Donohue III, "The Latest Misfires in Support of the 'More Guns, Less Crime' Hypothesis," 55 *Stanford L. Rev.* 1371 (2003).

182: **More questions about John Lott:** Tim Lambert has been tireless in researching and pursuing several Lott-related questions. http://timlambert.org/lott/. For a good summary of the Mary Rosh controversy see Julian Sanchez, "The Mystery of Mary Rosh," ReasonOnline, May 2003, http://www.reason.com/news/show/28771.html.

182: **Senator Craig cites Lott:** 146 Cong. Rec. S349 (daily ed. Feb. 7, 2000) (statement of Sen. Craig).

182: **Lott testifies:** Lott has testified before state legislatures in Nebraska (1997), Michigan (1998), Minnesota (1999), Ohio (2002), Wisconsin (2002), Hawaii (2000), and Utah (1999). On May 27, 1999, Lott testified before the House Judiciary Committee that the stricter gun regulations proposed by President Clinton either would have no effect or would actually cost lives, and a number of Republican members of Congress have since included favorable references in their speeches to Lott's work. Eighteen state attorneys general relied on "the empirical research of John Lott" in claiming: "There is an increasing amount of data available to support the claim that private gun ownership deters crime." See 145 Cong. Rec. H8645 (daily ed. Sept. 24, 1999) (statement of Rep. Doolittle); Letter from Bill Pryor, Att'y General of Alabama, et al., to John Ashcroft, U.S. Att'y General, July 8, 2002, http://www.nraila.org/media/misc/pryorlet.pdf; Nat'l Rifle Ass'n, Inst. Leg. Action, Right to Carry Fact Sheet 2007, http://www.nraila.org/Issues/factsheets/read.aspx?ID=18.

183: **Details on the Super Crunching analysis:** Lott started with a very simple model that only let the law have a one-time impact on crime. There's nothing wrong in starting with this specification. However, we found that, if you tested less constrained formulas that allowed the law to impact crime to varying degrees over time, his result often went away.

183: **Experts reject Lott's model:** Nat'l Acad. Sci., "Firearms and Violence: A Critical Review," http://www.nap.edu/books/0309091241/html/.

184: **Lott sues Levitt:** *Lott v. Levitt,* 2007 WL 92506, at *4 (N.D. Ill. Jan. 11, 2007).

184: **The offending paragraph from *Freakonomics*:** Steven D. Levitt and Stephen J. Dubner, *Freakonomics: A Rogue Economist Explores the Hidden Side of Everything* (2005), pp. 134–35. Lott also claimed he was defamed by a private email that Levitt sent to economist John McCall. See Complaint at 7, *Lott v. Levitt,* No. 06C 2007 (N.D. Ill. Apr. 10, 2006). ("It was not a peer refereed edition of the Journal. For $15,000 he was able to buy an issue and put in only work that supported him. My best friend was the editor and was outraged the press let Lott do this."). Lott maintains the special issue was peer-reviewed and that scholars with varying viewpoints were invited to contribute articles. As of this writing, the McCall email claim has not been dismissed.

184: **Levitt's endnote:** The endnotes also cited to one other article, Mark Duggan, "More Guns, More Crime," 109 *J. Polit. Econ.* 1086 (2001).

185: **Devil's advocacy:** Benedict XIV, *De Beat. et Canon. Sanctorum,* I, xviii (*On the Beatification and Canonization of Saints*).

185: **Devil's advocates in the boardroom:** Barry Nalebuff and Ian Ayres, *Why Not?: How to Use Everyday Ingenuity to Solve Problems Big and Small* (2003), pp. 8–9.

187–188: **Heckman's objections to randomized results:** James Heckman et al., "Accounting for Dropouts in Evaluations of Social Programs," 80 *Rev. Econ. and Stat.* 1 (1998); James J. Heckman and Jeffrey A. Smith, "Assessing the Case for Social Experiments," 9 *J. Econ. Perspectives* 85 (1995).

188–189: **How healthy are low-fat diets?:** Gina Kolata, "Maybe You're *Not* What You Eat," *N.Y. Times,* Feb. 14, 2006, p. F1; Gina Kolata, "Low-Fat Diet Does Not Cut Health Risks, Study Finds," *N.Y. Times,* Feb. 8, 2006, p. A1.

188–189: **WHI study on low-fat diets:** Women's Health Initiative homepage: http://www.nhlbi.nih.gov/whi/. See also B. V. Howard et al., "Low-Fat Dietary Pattern and Weight Change Over 7 Years: The Women's Health Initiative Dietary Modification Trial," 295 *JAMA* 39 (2006); B. V. Howard et al., "Low-Fat Dietary Pattern and Risk of Cardiovascular Disease: The Women's Health Initiative Randomized Controlled Dietary Modification Trial," 295 *JAMA* 655 (2006).

189: **Control vs. experimental low-fat diet groups:** Ross L. Prentice et al.,

"Low-Fat Dietary Pattern and Risk of Invasive Breast Cancer," 295 *JAMA* 629 (Feb. 8, 2006) ("Comparison group participants received a copy of *Nutrition and Your Health: Dietary Guidelines for Americans* and other health-related materials but were not asked to make dietary changes.").

189: **The Rolls-Royce of studies:** Gina Kolata, "Low-Fat Diet Does Not Cut Health Risks, Study Finds," *N.Y. Times*, Feb. 8, 2006, p. A1.

189–190: **Calcium supplement study:** R. D. Jackson et al., "Calcium Plus Vitamin D Supplementation and the Risk of Fractures," 354 *N. Engl. J. Med.* 669 (2006), erratum in 354 *N. Engl. J. Med.* 1102 (2006); 354 *N. Engl. J. Med.* 2285 (2006); Gina Kolata, "Big Study Finds No Clear Benefit of Calcium Pills," *N.Y. Times*, Feb. 16, 2006, p. A1.

190: **But is more information always better?:** Indeed, I have also pointed out contexts—for example, with regard to campaign finance contributions—where less information could be valuable. Bruce Ackerman and Ian Ayres, *Voting with Dollars: A New Paradigm for Campaign Finance* (2002).

CHAPTER 8

192: **Hiking estimates:** Ian Ayres, Antonia Ayres-Brown, and Henry Ayres-Brown, "Seeing Significance: Is the 95% Probability Range Easier to Perceive?" *Chance Magazine* (forthcoming 2007). It turns out that Sleeping Giant has been a wellspring of publication ideas. See also Ian Ayres and Barry Nalebuff, "Environmental Atonement," *Forbes*, Dec. 25, 2006. ("Ian was hiking at Sleeping Giant Park outside New Haven. As he exited the car, a tissue fell out of his pocket. Ian stooped to pick up the litter, but a gust of wind blew it a few feet out of reach. He walked over to pick it up and again the tissue skittered beyond his grasp.")

193: **Unlike the standard deviation, the variance is not our friend:** The variance is the other traditional measure of dispersion. The two concepts are closely related. The variance is simply the standard deviation squared. But the variance is not your friend. It is not at all intuitive. You can see the difference if you just look at the units in which the two are expressed. If I tell you that the average kid in Anna's class reads ten books in a month, you know what I mean. And very soon, you'll understand what it means to say that the standard deviation is three books. However, we'll all be long gone before we ever have an intuition for the fact that the variance is nine "squared books." So if you're ever told what some variance is, you should immediately turn it into a standard deviation (by taking its square root) and never, I repeat never, think about the variance again.

194: **Estimating a standard deviation from your intuitions (the answer):** Once you have your range of heights, you're ready to apply the Two Standard Deviation Rule in reverse. You see, because 95 percent of men fall within two standard deviations (both above and below) the average height, the distance between your upper and lower height represents four standard deviations. I have done this exercise dozens of time in class and most students say that 95 percent of men fall between 5'3" and 6'3", or 5'9" ±6 inches. The Two Standard Deviation Rule applied to this height range says that the standard deviation for male height is probably close to 3 inches. Of course this is just a rough estimate, but we should be pretty confident that the true standard deviation is not 5 inches or 1 inch.

195: **The man of the future is the man of statistics:** Oliver Wendell Holmes, Jr., "The Path of the Law," 10 *Harv. L. Rev.* 457 (1897).

196–200: **Point shaving in college basketball:** Justin Wolfers, "Point Shaving: Corruption in NCAA Basketball," 96 *Am. Econ. Rev.* 279 (2006); David Leonhardt, "Sad Suspicions About Scores in Basketball," *N.Y. Times*, Mar. 8, 2006, p. C1; David Leonhardt, "The NCAA's Response," *N.Y. Times*, Mar. 7, 2006 ("a player on roughly one out of every five teams has direct knowledge of point shaving.").

197: **Approximately normal distributions:** Male height and IQ scores are close, but not perfectly normal because the mathematical normal distribution has an infinitesimal chance of having outcomes close to positive or negative infinity. And we know there are no such things as negative heights or IQ scores.

The biggest problem with my knee-jerk application of the Two Standard Deviation Rule is that unlike the normal bell curve, many real-world distributions are skewed. Instead of being a symmetric bell shape on either side of the mean, a distribution will have a greater chance of taking on either a larger or a smaller number. The Two Standard Deviation Rule doesn't work as well for skewed distributions. There may not be a 95 percent chance that a variable will be within two standard deviations. Instead, the most that statistics can tell us is that there is at least a 75 percent chance that a non-normal random variable will be within two standard deviations of its mean. Nevertheless, I still apply the 2SD rule as a starting point whenever I'm trying to figure out the intuitive variation in some process.

199: **Justin Wolfers, superstar:** David Leonhardt, "The Future of Economics Isn't So Dismal," *N.Y. Times*, Jan. 10, 2007.

204: **Amar works backward:** One of this country's greatest scholars on the Constitution, Akhil Amar, used a similar method to back out an estimate of grade inflation. Yale doesn't release its grade statistics, so Akhil couldn't directly calculate the average GPA. However, it does release what grades a student needs in a particular year in order to qualify for the limited number of honors. The cutoff last year for magna cum laude (which is awarded to the top 15 percent of the class) was

3.82, while the cutoff for cum laude honors (which is awarded to the top 30 percent of students) was 3.72. Akhil realized that these two cutoffs were enough for him to estimate the average GPA. If he just assumed that the grade distribution at Yale is approximately normal, he could work backward from these two probabilities to estimate underlying mean and standard deviation. Here's what he said in an email: "If only 30 percent of students have GPAs of 3.72 or higher (approx .5 standard deviation above the mean) and if only 15 percent have GPAs of 3.82 or higher (approximately one standard deviation above the mean), then simple algebra suggests that the standard deviation is roughly .2 and the mean is roughly 3.62." [The "simple" algebra is that the two cutoffs for honors implicitly give us two equations with two unknowns. The two equations are: 5*sd+mean=3.72 and 1*sd+mean=3.82, where sd is the standard deviation and mean is the average GPA that we're trying to solve for. By solving the first equation for sd, and then substituting that solution into the second equation (remember "substituting out" from seventh-grade algebra!), we are able to solve for the mean: mean=2*3.72-3.82=3.62.] Of course, this is just an approximation, because the grades may not in fact follow the normal distribution. Still, Akhil's back-of-the-envelope estimate jibed pretty closely with the student newspaper's own survey of seniors, which suggested that the median GPA was between 3.6 and 3.7. Kanya Balakrishna and Jessica Marsden, "Poll Suggests Grade Inflation," *Yale Daily News*, Oct. 4, 2006, http://www.yaledailynews.com/articles/view/18226.

204: **The claim that women are "innately deficient":** Cornelia Dean, "Women in Science: The Battle Moves to the Trenches," *N.Y. Times*, Dec. 19, 2006.

204–205: **The claim that women lack "intrinsic aptitude":** Sara Rimer and Alan Finder, "After 371 Years, Harvard Plans to Name First Female President," *N.Y. Times*, Feb. 10, 2007.

205: **Summers's controversial presidency:** Newspaper reports have pointed to several other reasons why Summers may have felt pressure to resign. Slate columnist James Traub claims, "Summers was forced out of Harvard because he behaved so boorishly that he provided a bottomless supply of ammunition to his enemies." James Traub, "School of Hard Knocks: What President Summers Never Learned About Harvard," Slate, Feb. 22, 2006, http://www.slate.com/id/2136778/. *Washington Post* columnist Eugene Robinson said, "Summers is being forced to resign because, as brilliant as he is—and you don't become a tenured Harvard professor at twenty-eight, as Summers did, unless you're ridiculously brilliant—he proved to be a terrible politician." Eugene Robinson, "The Subject Larry Summers Failed," *Wash. Post,* Feb. 24, 2006. Even other parts of his speech about women in science rankled. For example, he proposed that another reason for the dearth of women in science was the relative unwillingness of women "to have a job that they think about eighty hours a week." Summers also analogized the shortfall to a rather bizarre set of comparisons: "It is after all not the case that the role of women in science is the only example of a group that is significantly underrepresented in an important activity and whose underrepresentation contributes to a shortage of role models for others who are considering being in that group. To take a set of diverse examples, the data will, I am confident, reveal that Catholics are substantially underrepresented in investment banking, which is an enormously high-paying profession in our society; that white men are very substantially underrepresented in the National Basketball Association; and that Jews are very substantially underrepresented in farming and in agriculture." Lawrence H. Summers, Remarks at NBER Conference on Diversifying the Science and Engineering Workforce, Jan. 14, 2005, http://www.president.harvard.edu/speeches/2005/nber.html.

206: **Studies finding sex differences in IQ standard deviations:** Ian J. Deary et al., "Population Sex Differences in IQ at Age 11: The Scottish Mental Survey 1932," 31 *Intelligence* 533 (2003). See also Ian J. Deary et al., "Brother—Sister Differences in the G Factor in Intelligence: Analysis of Full, Opposite-Sex Siblings from the NLSY1979," *Intelligence* (working paper, 2007).

208: **Possible flaws with Summers's methodology:** The difference in standard deviations may not be as great as 20 percent and the bell curves of intelligence may not follow a normal distribution—especially when you go so far into the tail.

209: **The Naegele rule:** Janelle Durham, "Calculating Due Dates and the Impact of Mistaken Estimates of Gestational Age," Jan. 2002, http://www.transitionto parenthood.com/ttp/birthed/duedatespaper.htm.

209: **Other methods of predicting due dates:** R. Mittendorf et al., "The Length of Uncomplicated Human Gestation," 341 *N. Engl. J. Med.* 461 (1999). For more on probabilities and predicting pregnancy terms, see W. Casscells et al., "Interpretation by Physicians of Clinical Laboratory Results," 299 *N. Engl. J. Med.* 999 (1978); David M. Eddy and Jacquis Casher, "Usefully Interpreting the Triple Screen Assay to Detect Birth Defects," working paper, Dept. of Statistics and of Biostatistics and Medical Informatics, Aug. 3, 2001; "Probabilistic Reasoning in Clinical Medicine: Problems and Opportunities," in *Judgment Under Uncertainty: Heuristics and Biases* (D. Kahneman et al., eds., 1982); Gerd Gigerenzer, "Ecological Intelligence: An Adaptation for Frequencies," in *The Evolution of Mind* (D. D. Cummins and C. Allen, eds., 1998), pp. 9–29; David J. Weiss, "You're Not the Only One Who Is Confused About Probability...," http://instructional1.calstatela.edu/dweiss/Psy302/Confusion.htm.

210: **Predicting contribution to the competitive bottom line:** Alan Schwarz,

"Game Theory Posits Measure of Baseball Players' Value," *N.Y. Times*, Nov. 7, 2004.

210–215: **Predicting the probability of Down syndrome:** N. J. Wald et al., "Integrated Screening for Down's Syndrome Based on Tests Performed During the First and Second Trimesters," 341 *N. Engl. J. Med.* 1935 (1999); Women's Health Information, Down syndrome, http://www.womens-health.co.uk/downs.asp; Miriam Kuppermann et al., "Preferences of Women Facing a Prenatal Diagnostic Choice: Long-Term Outcomes Matter Most," 19 *Prenat. Diag.* 711 (1999).

211: **The Bayes' theorem:** Thomas Bayes, "An Essay Towards Solving a Problem in the Doctrine of Chances," 53 *Phil. Trans.* 370 (1763). See also Eliezer Yudowsky, "An Intuitive Explanation of Bayesian Reasoning," 2003, http://yudkowsky.net/bayes/bayes.html; Gerd Gigerenzer and Ulrich Hoffrage, "How to Improve Bayesian Reasoning Without Instruction: Frequency Formats," 102 *Psych. Rev.* 684 (1995).

212: **Quantifying the probability of Down syndrome:** The tendency of some insurance companies to cover amniocentesis if the probability of Down syndrome is greater than the probability of miscarriage has a more sinister foundation. Instead of promoting their patient's welfare, the miscarriage probability rule minimizes the insurance companies' expected "cost-per-case detected."

214: **Bayes' theorem can be stated:** The posterior probability of cancer given a positive test is equal to the prior probability of cancer multiplied by a likelihood ratio, where the likelihood ratio is the probability of a positive test given that person has cancer divided by the probability of a positive test. Applied to these facts, the likelihood ratio equals 7.5 because the probability of cancer given a positive test is .8 and the probability of a positive test is .107 (107/1000). So Bayes' theorem says that we update the prior probability of 1 percent by multiplying by a likelihood ratio of 7.5 to yield the posterior probability of 7.5 percent.

215: **Suggested readings:**

Ray C. Fair, *Predicting Presidential Elections and Other Things* (2002).

Steven Levitt and Stephen J. Dubner, *Freakonomics: A Rogue Economist Explores the Hidden Side of Everything* (2005).

John Allen Paulos, *Innumeracy: Mathematical Illiteracy and Its Consequences* (1989).

John Donohue, *Beautiful Models, and Other Threats to Life, Law, and Truth* (forthcoming).